Smart Mobility and Intelligent Transportation Systems for Commercial and Hazardous Vehicles

Smart Mobility and Intelligent Transportation Systems for Commercial and Hazardous Vehicles

Edited by

T. Poongodi, PhD

Naga Swetha Pasupuleti, PhD

Balamurugan Balusamy, PhD

AAP | APPLE ACADEMIC PRESS

First edition published 2025

Apple Academic Press Inc.
1265 Goldenrod Circle, NE,
Palm Bay, FL 32905 USA

760 Laurentian Drive, Unit 19,
Burlington, ON L7N 0A4, CANADA

CRC Press
2385 NW Executive Center Drive,
Suite 320, Boca Raton FL 33431

4 Park Square, Milton Park,
Abingdon, Oxon, OX14 4RN UK

Library and Archives Canada Cataloguing in Publication

CIP data on file with Canada Library and Archives

Library of Congress Cataloging-in-Publication Data

CIP data on file with US Library of Congress

ISBN: 978-1-77491-538-7 (hbk)
ISBN: 978-1-77491-539-4 (pbk)
ISBN: 978-1-03268-404-8 (ebk)

About the Editors

T. Poongodi, PhD
Professor, School of Computing Science & Engineering at Galgotias University, Delhi– NCR, India

T. Poongodi, PhD, is currently working as a Professor in the School of Computing Science & Engineering at Galgotias University, Delhi–NCR, India. She was awarded her PhD degree in Information Technology (Information and Communication Engineering) from Anna University, Tamil Nadu, India. Her current research interests include network security, wireless ad hoc and sensor networks, Internet of Things (IoT), data science, and blockchain technology for emerging communication networks. She is CISCO, Oracle Academy certified, and teaches courses on networking and structured query language. Dr. Poongodi is the author of over 50+ (Scopus indexed) book chapters including in books published by some reputed publishers such as Springer, Elsevier, IET, Wiley, De-Gruyter, CRC Press, IGI Global. She has also published more than 30 (SCI/Scopus) papers in international journals and conferences. She has published more than 15 authored/edited books in the areas of Internet of Things, data analytics, blockchain technology, artificial intelligence, machine learning, and healthcare informatics, published by reputed publishers such as Springer, IET, Wiley, CRC Taylor & Francis, and Apple Academic Press. She adopts a universal and humanistic approach in her academic and research works. In her research, she has undertaken meticulous scientific studies of emerging issues in networking disciplines. She has many years of academic work experience in teaching and multidisciplinary research. She has received many awards, namely the Research and Innovation Award (2019, 2020, 2021) and Excellence in the Area of Research & Innovation/Academic Excellence/Extension Activities (2018–19, 2019–20) from Galgotias University. Prof. Dr. T. Poongodi has also received invitations to address international conferences as a keynote speaker and session chair. She is a senior member of the Institute of Electrical and Electronics Engineers (IEEE), IEEE Women in Engineering (WIE), and is an active member in various professional bodies, such as the Association for Computing Machinery (ACM), the International Association of Engineers (IAENG), the Institute of Research Engineers and Doctors (IRED), and International Association of Computer Science and Information Technology (IACSIT). She is the reviewer for international journals and conferences. She also has five Indian patents.

Naga Swetha Pasupuleti, PhD

Associate Director, International Relations at SRM University, AP, India

Naga Swetha Pasupuleti, PhD, is an Associate Director of International Relations at SRM University-AP. Dr. Swetha is a proficient academician and administrator with over 15 years of leadership experience in academics, research, industry, and administration. She has worked in various parts of the world (USA, Africa, UAE, etc.). As a researcher, her areas of specialization include sustainable transportation, transportation safety, ITS, and transportation planning; She has nine books and several publications in reputed national and international journals to her credit. She is a member of the Institute of Transport Engineers (ITE), USA; the Institution of Civil Engineers (ICE), UK; and the American Society of Civil Engineers (ASCE). As the Associate Director of International Relations. Dr. Swetha's remarkable versatility, experience, and expertise are an asset in enhancing the overseas visibility and engagement of SRM University-AP. After obtaining her master's in Civil Engineering and PhD from the University of North Carolina at Charlotte, North Carolina, USA, Dr. Swetha did her Post-Doc in IIT (Madras).

Balamurugan Balusamy, PhD

Associate Dean-Student Engagement, Shiv Nadar Institution of Eminence, Delhi-NCR, India.Adjunct Professor, Department of Computer Science & Information Engineering, Taylor University, Malaysia

Balamurugan Balusamy, PhD, is currently working as an Associate Dean-Student Engagement at the Shiv Nadar Institution of Eminence, Delhi-NCR,

India. He is also an Adjunct Professor in the Department of Computer Science and Information Engineering at Taylor University, Malaysia. Prior to this assignment, he was a Professor in the School of Computing Sciences and Engineering and Director of International Relations at Galgotias University, Greater Noida, India. Prof. Balusamy also served in the position of Associate Professor during his stint of 15 years of experience with VIT University, Vellore, India. He is included on Stanford University's list of Top 2% Scientists Worldwide 2023 in Data Science/AI/ML. His contributions focus on engineering education, blockchain, and data sciences. His academic degrees and 12 years of experience working as a faculty at global universities such as VIT University, Vellore, have made him more receptive and prominent in his domain. His passion is teaching, and he adapts different design thinking principles while delivering his lectures. He has over 200 papers published in high-impact journals published by Springer, Elsevier, and IEEE. He has also published over 200 edited and authored books and collaborated with eminent professors across the world from top QS ranked universities. He has visited over 15 countries for his technical coursework. He has attached several top-notch conferences. He serves on the expert advisory panel for several start-ups and forums and does consultancy work for the industry on industrial IoT. He has given over 200 talks at various events and symposiums.

Contents

Contributors

T. Lucia Agnes Beena
Department of Information Technology, St. Joseph's College, Tiruchirappalli, Tamil Nadu, India

S. Chidambaranathan
Vice Principal, St. Xavier's College (Autonomous), Palayamkottai, Tamil Nadu, India

Jagjit Singh Dhatterwal
Department of Artificial Intelligence and Data Science, Koneru Lakshmaiah Education Foundation, Vaddeswaram, Andhra Pradesh, India; Department of Computer Science and Applications, PDM University, Bahadurgarh, Haryana, India

Aditi Laxmi Giri
School of Computing Science and Engineering, Galgotias University, Greater Noida, Uttar Pradesh, India

Vaishali Gupta
Assistant Professor, School of Computing Science and Engineering, Galgotias University, Delhi NCR, India

S. Janarthanan
Assistant Professor, School of Computing Science and Engineering, Galgotias University, Delhi NCR, India

A. John
School of Computing Science and Engineering, Galgotias University, Greater Noida, Uttar Pradesh, India

Kuldeep Singh Kaswan
School of Computing Science and Engineering, Galgotias University, Greater Noida, Uttar Pradesh, India

Supriya Khaitan
Assistant Professor, Pillai College of Engineering, University of Mumbai, Mumbai, Maharashtra, India

M. Kiruthika
Assistant Professor, Department of Computational Intelligence, School of Computing, SRM Institute of Science and Technology, Kattankulathur, Chengalpattu, Chennai, Tamilnadu, India

Avadhesh Kumar
School of Computing Science and Engineering, Galgotias University, Greater Noida, Uttar Pradesh, India

T. Ganesh Kumar
Associate Professor, School of Computing Science and Engineering, Galgotias University, Delhi NCR, India

Vishwanadham Mandala
Enterprise Architect, 1311 Horse Shoe Bend, Greenwood, Indiana, USA

L. Murali
Associate Professor, Department of ECE, P.A. College of Engineering and Technology, Pollachi, Tamil Nadu, India

Pawan Kumar Singh Nain
Department of Mechanical Engineering, Galgotias University, Greater Noida, Uttar Pradesh, India

Nilanjana Pradhan
Research Scholar, School of Computing Science and Engineering, Galgotias University, Delhi-NCR, India

Sanjeev Kumar Prasad
Professor, School of Computing Science and Engineering, Galgotias University, Delhi NCR, India

Preety
Dr. Akhilesh Das Gupta Institute of Technology and Management, New Delhi, India

Priya Prowal
PhD Scholar, School of Computer Science and Engineering, Galgotias University, Greater Noida, Uttar Pradesh, India

Rajbala
Department of Mechanical Engineering, Galgotias University, Greater Noida, Uttar Pradesh, India

E. Rajesh
School of Computing Science and Engineering, Galgotias University, Greater Noida, Uttar Pradesh, India

S. R. Ramya
Department of Computer Science and Engineering, Dr. NGP Institute of Technology, Coimbatore, Tamil Nadu, India

Shrddha Sagar
Professor, School of Computing Science and Engineering, Galgotias University, Greater Noida, Uttar Pradesh, India

Aradhna Saini
Assistant Professor, G. L. Bajaj Institute of Engineering and Management, Greater Noida, UP, India

Ajay Shankar Singh
Professor, School of Computer Science and Engineering, Galgotias University, Greater Noida, Uttar Pradesh, India

E. P. Sumesh
Professor, School of Electronics Engineering (SENSE), VIT-AP University, Andhra Pradesh, India; Department of Electronics Communication Engineering, Middle East College, Muscat, Oman

R. Vidhyalavanya
Department of Computing and Electronics Engineering, Middle East College, Muscat, Oman

Ashok Yadav
School of Computing Science and Engineering, Galgotias University, Greater Noida, Uttar Pradesh, India

V. Yathavaraj
Assistant Professor, Department of Computer Science and Engineering,
Dr. NGP Institute of Technology, Coimbatore, Tamil Nadu, India

M. Yuvaraja
Associate Professor, Department of ECE, P.A. College of Engineering and Technology, Pollachi, Tamil Nadu, India

Abbreviations

ABSE	agent-based software engineering
ACO	ant colony optimizing
AGV	automated guided vehicles
AI	artificial intelligence
AIS	artificial immune system
AML	acute myeloid leukemia
ANN	artificial neural networks
AR	augmented reality
AVI	automatic vehicle identifiers
AVL	automatic vehicle locators
AVs	autonomous vehicles
BCO	bee colony optimization
BERT	bidirectional encoder representations from transformers
CAILM	continuous air interface long and medium range
CAV	commercial agricultural vehicles
CAVs	connected and autonomous vehicles
CB	customized bus
CNN	convolutional neural network
CPS	cyber-physical systems
CVD	cardiovascular disease
DBM	deep Boltzmann machines
DBMS	data stored in database management systems
DL	deep learning
DLT	distributed ledger technology
DNN	deep neural network
DSRC	dedicated short-range communication
EMMPF	efficient multiple model particle filter
ERP	enterprise resource planning
FastGCRNN	fast graph convolution recurrent neural network
FC	fully connected
FIVE	Ford immersive vehicle environment
FLM	fuzzy logic model

FN	false negative
FP	false positive
GA	genetic algorithm
GDPR	general data protection regulation
GPS	global positioning system
GRU	gated recurrent unit
HMI	human-machine interface
H-VEX	Hyundai vest exoskeleton
ICT	information and communication technology
IDMS	industrial data management system
IIoT	industrial internet of things
IoMT	internet of manufacturing things
IoT	internet of things
IoV	internet of vehicles
IT	information technology
ITS	intelligent transportation system
KNN	K-nearest neighbors
KYC	know-your-customer
LDA	latent Dirichlet allocation
LRM	logistic regression model
LSTM	long–short-term memory
M2M	machine-to-machine communication
MaaS	mobility-as-a-service
MAVs	manned aerial vehicles
MES	manufacturing execution system
MIT	Massachusetts Institute of Technology
ML	machine learning
MLP	multi-layer perceptron
MTLDNN	multitask learning deep neural networks
NB	Naive Bayes
NDP	network design problem
NLP	natural language processing
NNs	neural networks
OCC	optical camera communication
OPC-UA	open platform communication unified architecture
OT	operational technology
OWC	optical wireless communication
PATS	personal aerial transport systems

PAVs	personal aerial vehicles
PBS	pattern-based framework
PI	physical interface
PLC	programmable logic controller
PLM	product lifecycle management
PNN	probabilistic neural network
PReDicT	predicting response to depression treatment
PReLU	parametric rectified linear units
PSO	particle swarm optimization
PV	photovoltaic
PWPS	photovoltaic water pumping system
QMS	quality management system
RDBMS	relational database management systems
ReLU	rectified linear units
RF	random forest
RNN	recursive neural network
RP	revealed preference
RTU	remote terminal units
SA	search algorithm
SA	swarm optimization
SAE	Society of Automotive Engineers International
SCADA	supervisory control and data acquisition
SCM	supply chain management
SCO	supply chain orientation
SDMSM	secured data management system for smart manufacturing
SMS	smart manufacturing system
SP	scheduling planning
SP	stated preference
SQL	structured query language
SVM	support vector machine
TfL	transport for London
TN	true negative
TOD	transport-oriented developments
TP	true positive
UDRs	usage detail records
UFES	University of Espirito Santo
UITS	Urban Intelligent Transportation Systems
V2C	vehicle to cloud

V2I	vehicle to infrastructure
V2P	vehicle to pedestrians
V2R	vehicle to roadside unit
V2S	vehicle to sensors
V2V	vehicle-to-vehicle
VR	virtual reality
WMS	warehouse management system

Preface

Smart mobility envelops numerous components, including traffic-lessening measures like ride-sharing and bicycle sharing; remote correspondences, continuous information investigation, and AI utilized in self-ruling vehicles; and physical frameworks like parking spots and traffic signals. Urban areas use sensors, information stages, and programming to deal with their vehicle foundation and administrations as a solitary, composed framework. In commercial vehicle operations (CVO), intelligent transportation system (ITS) is utilized in business vehicle tasks to help improve the business vehicle security and administrative procedures and upgrade the effectiveness by consistently rethinking and altering the activities, both of the motor transporter industry and the state-run motor bearer offices.

Different segments of CVO include cargo organization, business vehicle regulatory procedures, roadside CVO security, onboard security checking, and hazardous material transportation. Smart mobility can give incredible key and strategic apparatuses for transport organizations, which, if appropriately applied and utilized, could acquire significant benefits advancing and fortifying their competitiveness. The utilization of such advancements has extraordinary implications for current commercial vehicle operations. Transportation engineering faces challenges of increasing complexity, such as providing safe, efficient, and reliable transportation while minimizing the impact on the environment and communities which has turned out to be quite difficult.

A constant increase in travel demand, fueled by economic development, and the ever-growing demands are some of those challenges that transportation professionals face including capacity problems, poor safety records, unreliability, environmental pollution, and wasted energy. Transportation systems are inherently complex systems involving a very large number of components and different parties, each having different and often conflicting objectives. It is found and evident that in recent years, there has been increased interest among both transportation researchers and practitioners in exploring the feasibility of applying artificial intelligence

(AI) paradigms to address some of the aforementioned problems in order to improve efficiency, safety, and environmental compatibility of transportation systems.

Introduction

Chapter 1: Smart Mobility and Intelligent Transportation Systems for Commercial and Hazardous Vehicles

In the view of rising population and rapid urbanization, a need for a smart transportation system that mitigates the issues of increasing traffic congestion such as wasted time, fatalities or pollution has become essential. Though attempts have been made to alleviate these issues by building more highways and apps to find alternate routes, they did not show much improvement to the current conditions. This chapter delves into the working and idea behind smart transportation systems and ITS for commercial and hazardous vehicles, and the integration of AI and ML into transportation systems which is believed to be the future of transport.

Chapter 2: Industrial IoT in Smart Manufacturing

Industrial IoT is being developed as a result of the Internet of Things (IoT) applications and the adoption of new technical developments in industrial systems (IIoT). IIoT primarily aims at improving the working quality of machines in an industrial environment. This setup allows for easy communication between machines that enhance the analysis quality and working, resulting in good data collection and economic benefits. This study explains the IIoT in smart manufacturing and industrial setting in a comprehensive manner covering the architecture, manufacturing, technologies involved, and challenges faced in the process.

Chapter 3: Learning Using Neural Networks for Smart Mobility

Since the emergence of neural networks, this technology has been integrated into various fields of study with the aim to simplify and automate the tasks at hand. Smart mobility, which primarily proposes to ease traffic issues, is a topic of high interest in the field of transportation and traffic management. Recent developments in the field of neural networks and their accuracy in decision-making have opened up various paths for integrating smart mobility into it. This chapter probes into the learning using neural networks in smart mobility starting with an overview of Deep neural networks (DNNs), a literature survey, and the application of various deep learning (DL) architectures to achieve the desired results.

Chapter 4: Smart Mobility Using AI and Intelligent Agents

Emerging technologies for a linked vehicle-infrastructure-pedestrian ecosystem, as well as big data, have made it simpler and less expensive to collect, store, analyze, exploit, and distribute multi-source data. Both artificial intelligence (AI) and mainstream computer science have embraced the concept of an agent. Anything that sees its surroundings acts independently to accomplish goals, and has the ability to learn and utilize information is considered an agent. The proposed study employs and reaffirms the use of artificial intelligence and intelligent agents in the ITS and smart mobility domain. The chapter's major goal is to outline the most significant theoretical and practical difficulties associated with creating intelligent agents and smart mobility systems that use AI.

Chapter 5: Blockchain for Efficient and Improved Strategy in Power Stations

With growing environmental concerns, the rush to develop sustainable methods of energy is higher than ever. The energy industry has generally been hesitant to adopt cutting-edge technology. Given how the current digital development strategy affects all players in the industry the transition to a new energy market would be necessary. This chapter proposes a novel approach that allows farmers to use an alternate method for irrigation using solar cells. This method also puts into service Blockchain technologies that allow the advancement of renewable energy.

Chapter 6: Artificial Intelligence-Based Smart Cities

Applications of AI are becoming more prevalent in everyday life. Given the revolutionary improvements it has brought to mobility, it is not an exaggeration to claim AI has revolutionary potential in the context of smart cities. Therefore, it is our role to document and offer comprehensive explanations of tomorrow's perplexing intelligent transport terms. By in-depth examining the under-researched convergence of AI, mobility, and the smart city and how this will change urban fates, this chapter provides a workable answer. It concentrates on important, intelligent transportation proposals like Connectivity and Highly Automated vehicles (CAVs), Fully Independent Private and Remotely Operated Aerial Vehicles (PAVs and UAVs), Maneuverability (MaaS), therapeutic principles like IoT technology, External Internet (PI), as well as more general changes like Industry 4.0. Finally, this research shows that the impact of green infrastructure goes much beyond the scope of cities themselves, having an impact on global standing, environmentalism, global warming, inclusivity, and unemployment.

Chapter 7: Optimized Machine Learning-Based Pedestrian Detection for Autonomous Vehicles

These days, driverless vehicle technologies are a developing trend. Important components to integrate into the autonomous vehicle system include lowering the percentage of pedestrian deaths pedestrian detection and collision avoidance systems. The identification of pedestrians has been a challenging and fascinating area of study for researchers who study intelligent transportation systems. This chapter presents a machine learning (ML) approach to identifying pedestrians in autonomous vehicles (AVs). The suggested research study focuses on optimization ways to get around the drawbacks of the classic ML approach by adopting a hybrid ML approach. The supplied classifier's detection accuracy has been increased by employing a hyperparameter optimization strategy similar to particle swarm optimization (PSO), which is utilized to improve the classifier's detection rate.

Chapter 8: Smart Railway Cargo Management System Using Big Data

With the massive influx of data coming in every second, the organization and association of data has become a mammoth task. To ease the burden of analyzing these complex data sets that are continually on the rise big data is used. Since its launch big data has found its way into many fields starting from cyber security to real-time traffic optimization systems. This section explores what is big data and consequently frames the idea of the impact of Industry 4.0 on transportation systems, leading to the objectives of the railway cargo management system using big data the need for its adoption into this sphere, and challenges encountered in the process.

Chapter 9: Transport in the New Era Using Artificial Intelligence

Technology has advanced significantly in the field of artificial intelligence since its creation in 1956. Though it was once perceived as a lot of abstract theory, it is now being used in a variety of areas, making things easier and optimal through automation. In this chapter, we have compiled an analysis of the issues with transportation-related technology. We examine how the use of AI-enabled technologies in the transportation industry would quickly relieve crowded areas, make traveling time more reliable for their clients, enhance the economies and reliability of their critical assets, and avoid inconvenience, and mishaps. We also proposed potential development areas for artificial intelligence in transportation.

Chapter 10: Machine Learning and IoT for Urban Intelligence

Life has been made easier by the invention of IoT, and it will soon become even simpler. Daily life is filled with several examples of IoT involvement. IoT was just recently introduced into the market, and it has a lot of potential to grow and play a significant part in many industries, including oil and gas, manufacturing, transportation, agriculture, and health care. This chapter focuses on IoT and machine learning-integrated healthcare applications, including improved robotic surgery. Machine learning contains several computations and factual models to determine acceptable outcomes based on data, so it will assist us in effectively achieving our specific aims.

CHAPTER 1

Smart Mobility and Intelligent Transportation Systems for Commercial and Hazardous Vehicles

S. JANARTHANAN, VAISHALI GUPTA, SANJEEV KUMAR PRASAD, and T. GANESH KUMAR

School of Computing Science and Engineering, Galgotias University, Delhi NCR, India

ABSTRACT

Smart mobility and intelligent transportation systems (ITS) for commercial and hazardous vehicles, at the beginning of the phase overview of commercial and hazardous vehicles and operations and the need for artificial intelligence (AI) and the internet of things (IoT) for intelligent mobility ITS. Slightly move on to introducing smart mobility and ITS user services primary architecture with components with commercial vehicle examples are various hazard classes and divisions of explosives, flammable gas, and non-flammable with different vehicle operations pickup trucks, semi-truck, vans, and taxis. An innovative user service, IoT is connected to provide real-time navigation and tracking system information. A significant role plays an essential vital function in intelligent traffic, public, safety, logistics with vehicle control, AI-based applications for the commercial functionality of autonomous trucking, railway cargo management services, and majorly integrating IoT in traffic level functionality. The same function is available as public transportation, and a case of studies of commercial services for author-reader interest, with the track it now, fleet services, and remote management. In conclusion, we intensely

Smart Mobility and Intelligent Transportation Systems for Commercial and Hazardous Vehicles.
T. Poongodi, Naga Swetha Pasupuleti, and Balamurugan Balusamy (Eds.)

studied privacy and security challenges in smart mobility. Threats like driverless cars, virtual reality (VR), and authentication and confidentiality with integrity used lightweight intrusion detection technology used for privacy protection.

1.1 INTRODUCTION

1.1.1 OVERVIEW OF COMMERCIAL AND HAZARDOUS VEHICLES

In the Commercial and Hazardous Vehicles for Transportation Act 1975, if not regularized or authorized properly, the most difficulty for public safety [1] and health-related issues will come in another was more number unnamed materials need to be regularized the flammable and explosives or in other form hazardous vehicles transmission (Figure 1.1).

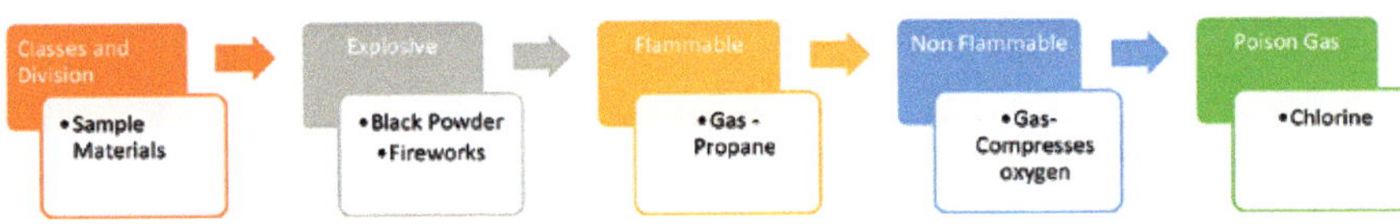

FIGURE 1.1 Hazard classes and divisions.

More than 2 billion to million transaction cargo was in the year of transportation shipments. In ways like a truck, air, water, and rail transmit a more significant number of hazardous materials, and classes were transferred every year and month and every day over the worldwide communities through a more substantial number of pathways above stated when we consider the petroleum products shipment around 40% of all the classes and divisions transmission of the ton of information over globally.

1.2 COMMERCIAL VEHICLE OPERATIONS

In general, commercial vehicle operation and transportation difficulties were much more complex to detecting the location of onboard items and objects was an important issue. Object analysis and detection of the geotagging location and position with detailed information related to those units [2]. On another end, heavy vehicle platform safety [3] commercial vehicle controller with actuator and wheel dynamics string stability controller and longitudinal tire model. These are more crucial in commercial vehicle operations (Figure 1.2).

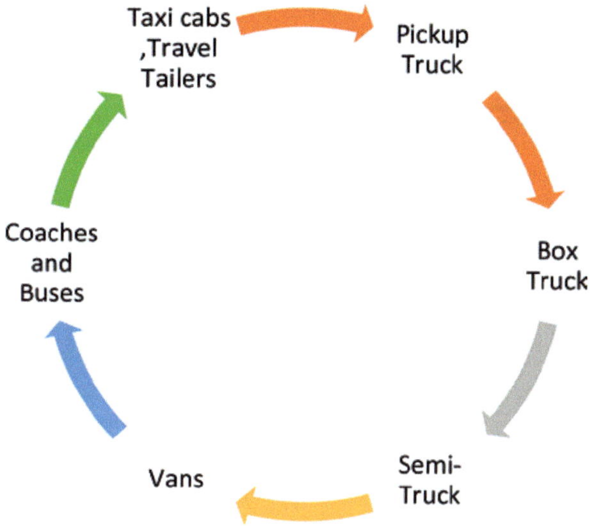

FIGURE 1.2 Commercial vehicle examples.

1.3 NEED FOR AI AND IOT FOR COMMERCIAL AND HAZARDOUS VEHICLES

Computer-based intelligence is successfully carried out in the auto esteem chain, including fabricating, planning, inventory network, creation, after creation, 'driver help' and 'driver risk appraisal' frameworks. In addition, AI has been proactively changing post-retail benefits like prescient upkeep and protection. Manufactured consciousness (AI) is a machine insight [4] device giving tremendous conceivable outcomes to the brilliant modern insurgency. Web of Things (IoT) is the saying of industry 4.0 transformation, including an overall framework for gathering and handling the information/data from capacity, activation, detecting, progressed administrations, and correspondence advancements. The mix of high velocity, strong, low-dormancy availability, and advances of AI and IoT will empower the change towards completely brilliant autonomous vehicles (AVs) that represent the integral between the natural world and computerized information for industry 4.0.

The reason for this chapter is to look at how the most recent methodologies in AI and IoT can aid the quest for AVs. It has been shown that human blunders are the wellspring of 90% of auto collisions, and the most secure drivers drive multiple times better compared to the standard (Wu et al. Mishap Analysis and Prevention. The robotized vehicle security is vast, and clients require multiple more modest satisfactory gamble levels. A portion of the unbelievable advantages of AVs is expanding vehicle security, a decrease in mishaps, a reduction in fuel utilization, delivery of driver time and business open doors, new potential market open doors, and diminished discharges and residue particles. In any case, AVs should utilize an enormous scope of information/data from their sensors and gadgets.

1.4 ITS AND SMART MOBILITY

A sophisticated software called an intelligent transportation system (ITS) aims to supply unique services connected to multiple modes of transport and traffic management, letting people be better efficient and aware of safer, better synchronized, and "smarter" use of mass transit. In a nutshell, smart mobility is a network of intelligent transportation and mobility. Smart mobility is a reimagining of the transportation infrastructure used in daily life and business that connects many components of technology and mobility. This includes not just traditional automobiles, electric vehicles, and public transit but also wholly new forms of mobility such as on-demand ridesharing services (Uber and Lyft) and carsharing programs.

1.4.1 ITS – AN INTRODUCTION

Consumer behavior shifts, such as the fall in private automobile ownership and the advent of wholly new mobility choices, are fast altering how people get about. Concerns about pollution, traffic congestion, productivity loss, and (of course) money have helped this concept gain popularity in the fleet business in recent years [5]. ITS are control and information that use incorporated communication systems and data management technologies to improve people and goods mobility, reduce traffic congestion, and effectively manage incidents, all while meeting transportation strategy aims and outcomes like demand governance and customer transportation prioritized metrics.

The phrase refers to a variety of techniques and tactics that can be deployed as standalone technical applications or by combining multiple systems to supply new (or enhanced) transportation services. In the context of road network operations, ITS provides individuals with the tools they need to change mobility and boost safety.

1.4.2 *ITS USER SERVICES*

The method gives legit information on a variety of topics, including the journey time, traverse speed, delays, traffic accidents, route changes, diversions, and work zone conditions [6]. This information is distributed via appropriate warning signs, motorway warning radios, online, texting, and autonomous cell phones (Figure 1.3).

FIGURE 1.3 SMART intelligent transportation system and its working process.

Major the impact of the working process automated system-level process for transportation best services to the user level communication and management system for access the internet services and mobile level alert system for import and export management services.

1.4.3 *ICT IN ITS*

On another end, ITS main task for ICT information and communication technology (ICT) management services, most importantly in all commercial and IoT transportation processes, which are connected with more devices for connection layer for the identification of the object and any devices much easy in nature. Internet connection with cloud services is more helpful for achieving this result. Components of ICT make the connection between the real-world entity to user-level interaction much easy in this level of management services [7]. In major making available of the software module along with hardware with the following transaction will be established with more number data sharing globally without internet connection information and communication management not working efficient manner. For best services availed through the cloud platform gives way for universal access to information with the best communication establishment around the globe.

In the new era digital world, one-to-one face level interaction moving from one location to another can communicate with each other. The other end enterprise-like business environment is most suitable for cost-cutting methods applied for effective management services using ICT, making available internet transactions via online shopping for medicinal and social media impact a lot. Significant challenges high-speed connectivity provided to the user with a secure level of the transaction also required for ICT management.

1.4.4 *ITS SYSTEM ARCHITECTURE AND COMPONENTS* (FIGURE 1.4)

FIGURE 1.4 ITS system architecture and components of ICT.

1.5 ROLE OF ARTIFICIAL INTELLIGENCE (AI) IN ITS

Artificial intelligence (AI) is an advanced emerging technology where machines perform like a human brain, including cognitive functions like perceiving, reasoning, learning, and problem-solving. The rapid breakthroughs in AI present limitless prospects to improve the performance of several public and private sectors, along with the transportation sector. The role of AI in the transport sector is to create and manage a sustainable transport system and overcome the challenges related to transport management, traffic congestion, and logistics faced by most cities across the world. The main aim of employing AI technology in the transport system is to provide safe, integrated AI-based decision-making, real-time traffic management, optimized mobility tools, and other efficient transportation network services.

Therefore, ITS is an embedded system consisting of leading-edge to create innovative transportation solutions using data and communication technology and traffic management systems. The use of AI and machine learning (ML) technologies make transportation management systems more intelligent by providing the ability to predict and make accurate predictions that help in real-time decision-making. Some of the critical functions of AI in ITS are intelligent traffic management, intelligent public transport, intelligent safety management, and intelligent logistics management (Figure 1.5).

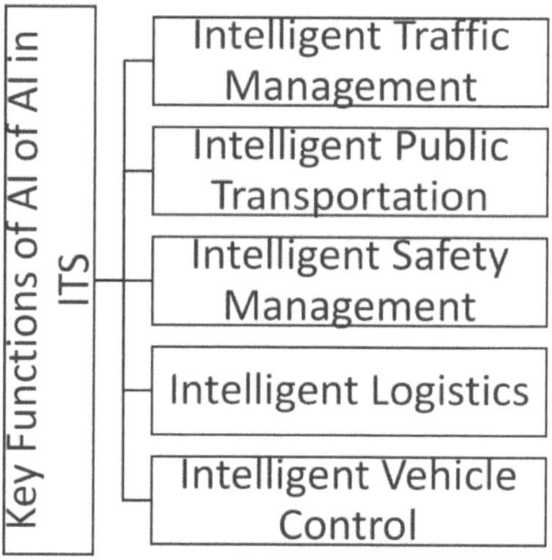

FIGURE 1.5 Key functions of AI in ITS.

With the help of data generated from the devices or sensors installed in the vehicles, AI provides solutions to transportation-related issues. AI offers solutions to traffic problems through real-time traffic light tracking or tracing to avoid traffic congestion by suggesting alternative routes during peak hour traffic time. This helps in traffic management. To manage public transport, AI helps predict the weather, traffic patterns, and road management by sending an alert to the on-duty officer so that drivers and commuters will receive information before the commencement of the trip. AI also provides solutions to [17] safety-related issues by predicting accident-prone areas based on road conditions, sending real-time alerts to drivers, etc. AI also offers intelligent logistics management by reducing manual paperwork like generating a large number of invoices and documents manually, sharing data across vehicles and routes, and by vehicle scheduling. It leads to a cost-effective system. AI also plays an essential role in lifting existing vehicle control systems. There are many AI-based algorithms incorporated in vehicles to reduce fuel or energy consumption, improve autonomous driving, or provide anti-lock advanced braking systems. Table 1.1 summarizes the transportation system-related challenges and the role of AI.

1.5.1 AI APPLICATIONS IN COMMERCIAL VEHICLES

A commercial vehicle can be classified as a vehicle whose primary purpose is to transport goods or materials rather than passengers. People can use a commercial vehicle for various purposes, both personal and commercial related. Examples of commercial vehicles include trucks, cargo vans, taxicabs, etc. One of the major areas of AI application is in the vehicle itself. AI plays an essential role in autonomous driving, environmental detection, control of the vehicle, and many other in-vehicle systems.

1.5.1.1 AUTONOMOUS TRUCKING

Trucking is the main target for AI interventions due to environmental challenges and emission regulations of transportation systems [8]. Independent trucks are a vehicle that utilize AI to robotize all that from delivery yard tasks to long stretch conveyances. Autonomous trucks are based on long-range, high-resolution sensors such as GPS, cameras, radar, etc., in

TABLE 1.1 Role of AI

Key Function	Challenges	Role of AI
Traffic management	Unpredictable traffic congestion. Routing. Increased cost due to traffic congestion. Peak hour congestion control. Increased number of vehicles on the road.	Machine learning algorithms to predict traffic build up. Alternative route suggestions. Identification of polluted air. Real-time tracking of congestion and AI-enabled traffic lights.
Public transport	Heterogeneous data from different sources. Traffic congestion. Delivery time and place variation. Damaged road.	Data patterns prediction through machine learning algorithms. Route optimization. Suggestions to improve driving patterns. Automatic road alerts.
Safety management	Fatigue and tiredness of drivers. Continuous driving hours and unknown terrain. Low performance and safety issues. Increased time and cost. Repair or refueling.	Auto-pilot system activation. Health monitoring of drivers. Blindspot alert, adaptive cruise control, advanced driver assistance systems. Optimization of routes. Remote control management.
Logistics management	Need for repair and maintenance. Increase in production and delivery cost. A large number of manuals entered invoices and documents. Anomalies in invoices. Compliance. Verification.	Combining data from IoT sensors, maintenance logs – prediction models are created. Connected vehicles scheduling predictive and preventive maintenance. Shared data across vehicles and routes. AI-based systems retrieve data with ease from the network. Prediction and tackling of fraud detection. Natural language processing technologies for interpretation of invoices.
Vehicle control management	Anti-lock breaking system. Vehicle's lateral and frontal control. The optimized vehicle suspension system. Automatic guided vehicle. Minimized consumption and emission in electric vehicles.	Genetic algorithm-based charging control system for electric vehicles. Expert systems to improve autonomous driving and advanced braking systems. ANN-based vehicle control system. NLP and computer vision for automatic vehicles.

combination with actuators, control units, and some deep neural networks (DNNs) to achieve elite execution energy-productive frameworks to further develop security and proficiency for ordinary coordinated factors. Some AI technologies are intended to take over certain functionalities while some are intended to completely replace the driver such as the human driver. Fully automated trucks are still in infancy but AI already influencing certain driving functions. It is more complex to test fully automated trucks in urban areas due to complicated road systems [18] and infrastructure. Truck platooning is an example where AI has been intervened. Truck platooning is a technique where the coupling of many heavy goods vehicles within a minimal distance of each other is possible and it also allowed vehicles to accelerate or brake automatically and simultaneously. Figure 1.6 illustrates truck platooning. The lead vehicle is driven by a human driver to avoid complex traffic situations such as roundabouts or unexpected incidents.

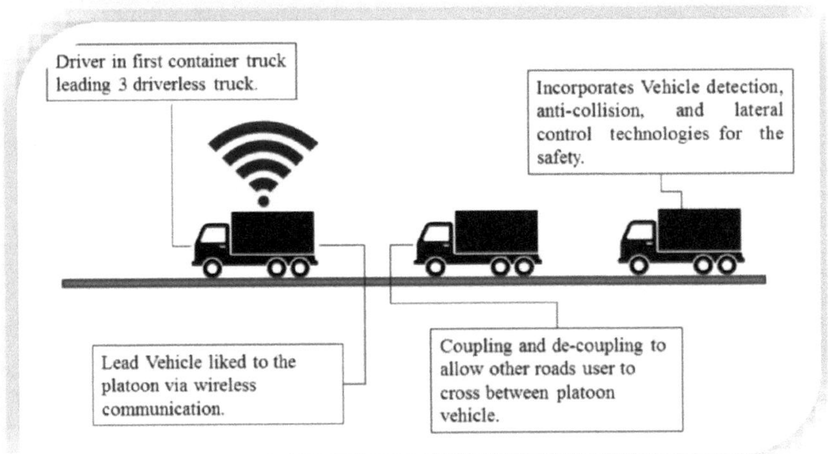

FIGURE 1.6 Truck platooning.

1.5.1.2 AI-ENABLED RAILWAY CARGO

To improve the efficiency of rail transportation solutions, GE transportation has created 'intelligent' locomotives. These brilliant cargo trains are outfitted with sensors and cameras that catch track, front and back.

The sensor's information is taken care of by machine-learned insightful calculations which total the information not too far off on the edge of the entryway, considering locally available ongoing independent direction.

1.5.2 ROLE OF IOT IN ITS

An ITS is an application that allows people to use a safer, secure, more synchronized, and smarter transportation system. Internet of things (IoT) plays an important role in ITS as it has great potential to transform transportation into a smart system by use of the physical network that consists of embedded sensors, actuators, and auto control networks that can gather real-time information in the network. To achieve effective communication between vehicles, IoT plays a major role in ITS.

Recently, IoT integrated ITS applications have gotten more attention and already they have entered their initial phases [9]. IoT technologies promise to build the transportation industry smarter and more successfully by reshaping transportation solutions to provide high safety, more efficient travel, improved vehicle control, and more strategic traffic management.

1.5.2.1 INTEGRATING TRAFFIC SYSTEM WITH IOT

The traffic system in ITS can be benefited from the integration of IoT in the normal operation of ITS. IoT stores sensor data and compute it using data analytics to manage and assist traffic system effectively. The various applications of IoT in traffic systems include traffic tracking and monitoring in case of traffic rule violation, traffic flow prediction, theft vehicle detection prioritized vehicle scheduling, and many more. With the use of data analytics techniques and multivariable phase space reconstruction methods, IoT helps in reducing traffic congestion. With the help of IoT, the connection between people and automobiles, roads, and vehicles can be established to resolve traffic congestion and air pollution problems to provide safe and secure transportation. Information can be accumulated from CCTV taking care of inside urban communities and move this vehicle-related information to traffic the executive's habitats to oversee shrewd leaving, traffic signals and to give savvy mishap help.

1.5.2.2 INTEGRATING VEHICLES WITH IOT

Nowadays many cars have Internet connectivity, sensors, and actuators to monitor everything from the brakes and motor to tire strain and fumes gas synthesis [10]. IoT can help connect automobiles in the future by using in-vehicle networks, radars, sensors, and cameras to identify and communicate with one another, preventing crashes, and facilitating traffic flow. IoT-powered Vehicle tracking systems help automobile companies to manage their fleets effectively. IoT-powered functionality in-vehicle systems include driver monitoring like driver idle time and driving style, trip scheduling, fleet tracking, vehicle load monitoring, speed, brake, and acceleration alerts, and calculation of distance traveled and fuel consumption.

1.5.2.3 INTEGRATING PUBLIC TRANSPORT SYSTEM WITH IOT

IoT technology is already widely used in the public transportation sector. There are several applications including integrated ticket systems, automated fare collection, traveler data show frameworks, and other progressed operations arrangements, that are IoT empowered and help social and financial issues in open transportation frameworks. IoT technology helps to connect public transport systems and helps in real-time vehicle tracking, real-time data analysis, and management, and to provide personalized travel information [11].

1. **Real-Time Vehicle Tracking:** Through cell phones and traveler data shows at travel pauses and stations, continuous vehicle tracking lets customers communicate better with public transportation companies and offer exact arrival and departure times.
2. **Information Examination and Continuous Administration:** It helps to monitor real-time progress and allows transportation agencies to make prior arrangements for unpredicted accidents, road works, emergencies, etc. Real-time data analysis helps to reroute the vehicle in case of emergencies.
3. **Personalized Travel Information:** Through the data collected from sensors and IoT devices, transport companies can track, and screen suburbanite conduct and travel designs and straightforwardly give customized data to their advanced cell or gadgets if

there should arise an occurrence of refreshed data, for example, delays, station terminations or re-directing.

1.5.3 IOT IN COMMERCIAL VEHICLES: CASE STUDIES

The transportation business has received the rewards of IoT innovation. The IoT is helping transportation organizations in planning the most effective courses and augmenting fuel use, coordinating operations organizations in following and following their merchandise, and stopping new companies from checking open spots continuously, among different applications.

In a commercial vehicle, the IoT means real-time data that cannot only help improve performance and safety but also helps to improve the efficiency of the manufacturing industries. IoT technology in commercial vehicles can help in better livestock monitoring, precision farming in the case of commercial agriculture vehicles, and virtually AVs.

1.5.3.1 TRAK IT NOW: IOT SOLUTION FOR COMMERCIAL AGRICULTURAL VEHICLES (CAVS)

The use of IoT technology in the agriculture sector not only means improving the lives of farmers in developed countries but also means real advancements for farmers in developing countries like India.

Trak it Now is a unique IoT solution that offers a connected vehicles platform to vehicle manufacturers to help in tracking the vehicles and also provides complete health information of the vehicle enabling them to do proactive maintenance. The insights from the vehicle data can be used as key inputs for their R&D activities.

With the help of new mobile technology, Trak it Now provides several options for farm vehicles including location tracking, engine health monitoring, air clogging, fuel level and oil pressure monitoring, and vehicle immobilization. The key to implementing IoT technology is identifying and creating sensor points that allow vehicle parameter data to be captured through Track it Now IoT devices. By connecting the Track, it Now IoT devices to the battery, they're able to send data to the central server at specified intervals, and in case of critical information about the health of the vehicle is identified; the information is passed to the customer's mobile device. The IoT technology also introduced a vital feature "Vehicle Immobilization" that helps the vehicle owners to take full control of the vehicle during situations like a vehicle being stolen (Figure 1.7).

FIGURE 1.7 IoT enabled commercial agricultural vehicle.

1.5.3.2 SCANIA TRUCKS: IOT-BASED COMMERCIAL VEHICLE AND FLEET SERVICES

Scania Trucks provides IoT solutions to truck fleet owners throughout the world to help them grow their businesses and revenues. Using IoT devices, connected trucks give large volumes of vital information in real-time. Scania Trucks' connected services are developed using this real-time data from connected trucks and provide services like lowering fuel consumption, increasing vehicle uptime, and more. Scania Trucks also provides a key service that boosts vehicle uptime and facilitates direct interaction with the workshop. Since the trucks are always connected, they can automatically report pending faults or key diagnostic data to workshop or fleet owners which will help to facilitate service planning with minimum downtime for fleet owners. The major goal is to develop more cost-effective IoT solutions that boost production, performance, and revenues while lowering emissions.

1.5.3.3 MAERSK: IOT-BASED REMOTE CONTAINER MANAGEMENT

The organization Maersk offers an Internet-of-Things-based far-off compartment the board stage that utilizes inside sensors to gather and communicate constant information on everything from temperature and stickiness to CO_2 levels in freight trucks. It additionally takes into consideration continuous day-in and day-out GPS holder following, programmed warnings that keep freight proprietors educated regarding any temperature or pulldown rate distortions, freight rerouting, and expanded security.

1.5.3.4 NEXT TRUCKING: IOT-BASED SERVICE PLATFORM FOR TRUCKS

NEXT TRUCKING is an IoT-based solution that connects truckers and shippers. These IoT administrations incorporate continuous shipment following and defer notices, support from a committed in-house account administrator, transporter task through the versatile application, transporter map pinpointing, coordinating with accessible delivery loads, and many more.

1.6 PRIVACY AND SECURITY CHALLENGES IN SMART MOBILITY

Smart mobility is an application of automation in ITS that provides users with real-time traffic information as well as alternate route ideas in the case of traffic congestion [12]. The automated ITS has completely transformed how people, goods, and services are transported from one place to another place. The IoT and the internet of vehicles (IoV) both play a role in smart mobility, but when objects are connected, security and privacy become key concerns. Smart mobility applications catch a huge measure of direction information that can be utilized to surmise a client's area and versatile ways of behaving. In brilliant versatility, security manages unlawful admittance to data. The interest in associated gadgets and the quick age of enormous measure of individual information raise genuine security concerns connected with information sharing among gadgets and clients. Also, the whole organization is available to outside assaults and breaks because of the associated gadgets. In addition, some reasons challenge privacy and security in smart mobility.

1.6.1 IOT BOTNETS IN IOT-BASED INTELLIGENT TRANSPORTATION SYSTEM (ITS)

An IoT botnet is a network of compromised IoT devices such as web cameras, routers, printers, wearables, and DVRs, infected with malware that allows the attacker to control them. These IoT botnets infect devices and spread the infection to many heterogeneous IoT devices that are connected in an IoT environment and cause a Distributed Denial of Service attack against target servers. Therefore, IoT botnets have raised serious security and privacy threats to IoT-based systems.

1.6.1.1 *THREATS OF DRIVERLESS CARS IN ITS*

Autonomous vehicles (AVs) aim to minimize traffic accidents and provide a smart transportation system. However, the growth of AVs can be seen as a major security concern. In the case of hacking AVs, life security and information protection will be undermined. Independent vehicle programmers can take advantage of safety bugs to direct remote goes after, for example, controlling breaks and guiding, closing down the motor, and so on. At the same time, personal data collected by vehicles can also be stolen that causing significant privacy issues.

1.6.1.2 *VIRTUAL REALITY (VR) DEVICES PRIVACY ISSUES*

In the smart technology-driven environment, virtual reality (VR) technology embraces various areas, but technology developers have not made suitable and complete protection contemplations. Major privacy issues include unencrypted communication channels between VR devices, sensitive and personal information shared with third parties, and threat to data stored by sensors.

1.6.1.3 *SECURITY AND PRIVACY THREATS POSED BY IN SMART MOBILITY*

In smart mobility applications, AI plays an indispensable role in AVs, traffic congestion control systems, real-time road alerts, and real-time traffic prediction. However, the expanding usage of AI raises security and privacy concerns, such as dangers to personal data that service providers employ to extract sensitive information that goes beyond their primary business purposes. Hackers that understand AI are also sophisticated and capable of attacking ML-based protective mechanisms to undermine ML training algorithms and reduce their reliability.

1.6.1.4 PRIVACY REQUIREMENTS AND SECURITY IN SMART MOBILITY

Considering the complex environment of smart mobility that includes many IoT devices and AI systems, there are some security and privacy requirements to secure ITS.

1.6.1.5 SECURITY AND AUTHENTICITY

Authentication, which ensures authorized access to services across a heterogeneous system, is a core requirement of a smart system. Smart mobility IoT devices must verify the network, other devices, and communication from management stations. In the case of explosively growing data in smart systems, there is a need to develop technologies that assure real-time and precise authentication.

In IoT based smart environment, to prevent the data or information from passive attacks or wrong exposure, confidentiality is important. Therefore, encryption-based technologies are important to prevent the confidentiality of information transmission between devices.

1.6.1.6 AVAILABILITY AND INTEGRITY

Availability ensures the availability of devices and services and maintains effective functioning while the system is under onslaught Because smart devices are vulnerable to assaults, a smart system must be able to identify abnormalities and be resilient enough to withstand defects and failures induced by attacks. The integrity of IoT devices and data shared between devices and the cloud must be assured in an IoT-based environment. In smart applications, data can easily tamper during transmission across many devices if the environment is not protected.

1.6.1.7 METHODS FOR DETECTING AND PREDICTING INTRUSIONS

To build a secure system, a smart system must detect the conditions to monitor the different operations with stipulated time duration. In heterogeneous and complex smart mobility environment, simple intrusion detection systems are found to be inflexible and unrealistic as they are used in anomaly detection, case of misuse detection, and specification-based detection [13]. Since sensors and smart devices are resources constrained, there is a need to develop lightweight intrusion detection systems. Prediction of incoming threats in advance is also a better solution to secure a smart system from attacks.

1.6.1.8 PRIVACY PROTECTION

In a smart mobility environment, the main causes of privacy breaches are malware in smart devices, packet interception in insecure communication, and leakage of sensitive data. According to a survey, data such as data that can be seen, data that has been repurposed, data that has been published, and data that has been leaked, are used to hack privacy. To protect privacy, Differential privacy technology, encryption methods, and anonymous procedures must all be used.

1.6.1.9 SMART MOBILITY – CHALLENGES AND SOLUTIONS

Smart mobility is a network [14] of intelligent transportation and mobility. Intelligent mobility is a reinvention of transportation infrastructure utilized in daily life and business that links several technological and mobility components. Current and future challenges in intelligent mobility, design the most effective, safe [15, 16], and safer transport systems for the public with mobility-as-a-service (MaaS). Make some innovations and adoption like autonomous, network-connected, shared vehicles. Create the most effective policies and strategies for quality life standards and measures [17, 18]. Create a public-private partnership to address air quality management, traffic, and various issues. Building stable and applicable infra for public and private sectors.

Urban mobility challenges the public's clean, secure, reliable, and connected autonomous vehicle innovation and adoption. Creating more suitable policies and strategies for public and private partnerships to address many issues and challenges.

1.7 CONCLUSION

Commercial and hazardous vehicle smart mobility and intelligent transportation. At the outset of the phase, we evaluated the overview of commercial and hazardous vehicle operations and the need for AI and IoT for intelligent mobility systems. Chapter 2 presents smart mobility and

ITS with a detailed look at the ITS user services basic architecture and components. Various danger classifications and divisions of explosives, flammable gas, and non-flammable with varied vehicle operations pickup trucks, semi-trucks, vans, and taxis are examples of commercial vehicles. IoT is a cutting-edge user service that connects to give real-time navigation and tracking system data. Intelligent traffic, public safety, logistics with vehicle control, AI-based apps for commercial functionality autonomous trucking, railway freight management services, and massive integration of IoT in traffic-level functionality all play a key role. With track it now, fleet services, and remote management, the same function is provided as public transportation, as stated in the case of studies commercial services for author-reader interest. Finally, we looked at the privacy and security issues that smart mobility presents. Lightweight intrusion detection technology was employed for privacy protection against threats such as autonomous automobiles, VR, authentication, and secrecy with integrity.

KEYWORDS

- **autonomous trucking**
- **commercial agricultural vehicles**
- **commercial vehicle operations**
- **intelligent transportation system**
- **privacy protection**
- **smart mobility**
- **virtual reality**

REFERENCES

1. For more information related to that. https://www.nap.edu/read/11198/chapter/4 (accessed on 2 November 2023).
2. Chen, C. H., Lin, J. H., Kuan, T. S., & Lo, K. R., (2014). A high-efficiency mobile positioning system by using commercial vehicle operation data based on cloud computing techniques. In: *2014 IEEE International Conference on Internet of Things (iThings), and IEEE Green Computing and Communications (GreenCom) and IEEE*

Cyber, Physical and Social Computing (CPSCom) (pp. 312–315). doi: 10.1109/iThings.2014.57.

3. Devika, K. B., Rohith, G., Shreya, Y. V. R., & Subramanian, S. C., (2020). A dynamics-based adaptive string stable controller for connected heavy road vehicle platoon safety. In: *IEEE Access* (Vol. 8, pp. 209886–209903). doi: 10.1109/ACCESS.2020.3039797.
4. Khayyam, H., Javadi, B., Jalili, M., & Jazar, R., (2020). *Artificial Intelligence and Internet of Things for Autonomous Vehicles*. 10.1007/978-3-030-18963-1_2.
5. For more information related to that https://rno-its.piarc.org/en/intelligent-transport-systems/what-its (accessed on 2 November 2023).
6. For More Information Related to that Refer the Link. https://www.geospatialworld.net/blogs/what-is-intelligent-transport-system-and-how-it-works/ (accessed on 2 November 2023).
7. For More Information Related to that Refer the Link. https://www.techtarget.com/searchcio/definition/ICT-information-and-communications-technology-or-technologies (accessed on 2 November 2023).
8. Ning, H., Hong, L., Jianhua, M., Laurence, T. Y., & Runhe, H., (2016). Cybermatics: Cyber–physical–social–thinking hyperspace-based science and technology. *Future Generation Computer Systems, 56*, 504–522.
9. Abduljabbar, R., et al., (2019). Applications of artificial intelligence in transport: An overview. *Sustainability, 11*(1), 189.
10. Iyer, L. S., (2021). AI enabled applications towards intelligent transportation. *Transportation Engineering, 5*, 100083.
11. Chand, H. V., & Karthikeyan, J., (2018). Survey on the role of IoT in intelligent transportation system. *Indonesian Journal of Electrical Engineering and Computer Science, 11*(3), 936–941.
12. Cui, L., et al., (2018). Security and privacy in smart cities: Challenges and opportunities. *IEEE Access, 6*, 46134–46145.
13. Paiva, S., et al., (2020). Privacy and security challenges in smart and sustainable mobility. *SN Applied Sciences, 2*, 1–10.
14. For more information related to that refer the link. https://hub.beesmart.city/en/solutions/smart-mobility/smart-mobility-challenges-and-solutions-in-smart-cities (accessed on 2 November 2023).
15. Vijayalakshmi, S., Gayathri, S. P., & Janarthanan, S., (2021). Blockchain security for artificial intelligence-based clinical decision support tool. In: *Internet of Things, Artificial Intelligence and Blockchain Technology* (pp. 209–240). Springer, Cham.
16. Janarthanan, S., & Rajan, K., (2017). *Secure Efficient Geometric Range Queries on Encrypted Spatial Data. 1*(1), 1–4 (NA).
17. Poongodi, T., & Karthikeyan, M., (2016). Localized secure routing architecture against cooperative black hole attack in mobile ad hoc networks. *Wireless Personal Communications, 90*(2), 1039–1050.
18. Poongodi, T., Khan, M. S., Patan, R., Gandomi, A. H., & Balusamy, B., (2019). Robust defense scheme against selective drop attack in wireless ad hoc networks. *IEEE Access, 7*, 18409–18419.

CHAPTER 2

Industrial IoT in Smart Manufacturing

T. LUCIA AGNES BEENA,[1] M. KIRUTHIKA,[2] and ARADHNA SAINI[3]

[1]*Department of Information Technology, St. Joseph's College, Tiruchirappalli, Tamil Nadu, India*

[2]*Assistant Professor, Department of Computational Intelligence, School of Computing, SRM Institute of Science and Technology, Kattankulathur, Chengalpattu, Chennai, Tamilnadu, India*

[3]*Assistant Professor, G. L. Bajaj Institute of Engineering and Management, Greater Noida, UP, India*

ABSTRACT

The industrial internet of things (IIoT) is one of the vital technologies that adds elegance to the industries. The smart manufacturing Industries that adopt IIoT include smart machines, and factories with a high level of intelligence embedded in them. The smart manufacturing industry has large scale networks with heterogeneous gadgets connected to produce dynamic states generating massive data. The networking system utilize internet that follows open standards for communication and cloud technologies to gain access to the devices securely. Compared with traditional manufacturing process, smart manufacturing is supposed to handle huge amount of data. The advanced analytical tools of Big Data are employed for processing and mobile technologies for faster communication. Adoption of IIoT ensures improved security, efficiency, profitability, better safety, performance with minimized carbon-di-oxide emission. The strength of IIoT depends on different technologies such as IoT, edge cloud computing, big data analytics, blockchain, artificial intelligence (AI), cyber physical systems

Smart Mobility and Intelligent Transportation Systems for Commercial and Hazardous Vehicles.
T. Poongodi, Naga Swetha Pasupuleti, and Balamurugan Balusamy (Eds.)

(CPS) and augmented and virtual reality (VR). A framework known as industrial data management system (IDMS) is discussed in this chapter, that collects real-time data, transmits the data, process the data and store the data. This efficient framework gives solutions for managing, transmitting, and securing data. The industrialists and engineers face numerous challenges such as security issues, IIoT integration, connectivity, and visibility of components involved in the system. This chapter discusses a portion of the science and innovation, challenges, like creating clever gadgets, building the organization climate, broad information investigation and preparing, and advanced creation for the better implementation of IIoT in smart manufacturing.

2.1 INTRODUCTION

Latest development in the field of Information and Communication Technologies, Computer Science and Manufacturing Industry leads to the Industry 4.0. The cyber physical systems (CPS) that relays on the Industry 4.0 converges the physical and virtual world in smart manufacturing [1]. Nowadays, the word 'smart' is closely associated with IoT. Manufacturing refers to industrial fabrication and in current scenario, it happens in computerized way with the aid of new technologies with proper supply and value chain. The smart manufacturing system (SMS) adopts the CPS for transparency, improved understanding, refined decision making and self-optimization of the systems. CPS enhance the capabilities of the SMS for the efficiency, productivity, and safety of the industries. The ultimate goal of the SMS is to increase profit, enrich customer experience by decreasing cost and to remain relevant in the market [2]. The key players that transform SMS are Sensors, Actuators, Transducers to capture data, Edge Computing for computing at the edge of the network, Cloud Computing for remote access and storage, big data for data analytics, artificial intelligence (AI) to improve decision making, internet of things (IoT) and Network technologies for man-to-machine, machine-to-machine communication, manufacturing execution system (MES) and enterprise resource planning (ERP).

Mohammed et al. [3] proposed a model on manufacturing processes and it is drafted in Figure 2.1. The computational intelligence was used to control the whole manufacturing processes. This model explains the manufacturing process integrated with computer by dividing the tasks into different layer. The different layers are business planning, operational management, monitoring, sensing, and Manipulating and Production process. Decision related to end product are resolved at the Business planning layer.

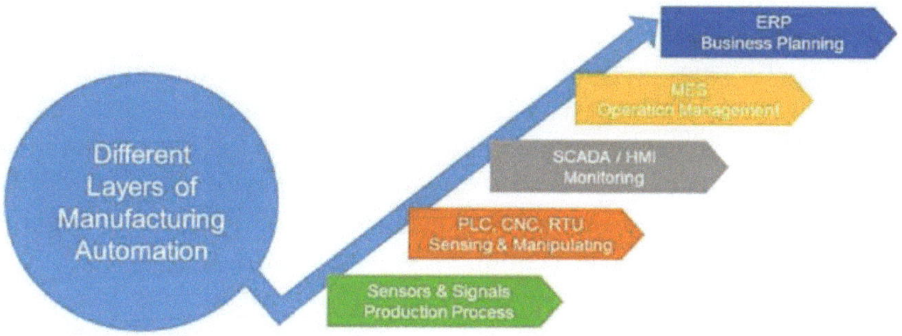

FIGURE 2.1 Different layers of manufacturing automation.

ERP is applied at the business planning layer. ERP ropes the information flow between the enterprise and the stakeholders outside the enterprise. Also, it efficiently handles the managerial issues. At the operational level, to face the varying demands, the shop floor data has to be appropriately utilized which is better handled by MES along with ERP. MES optimizes the overall manufacturing operations starting from work order to finished product by effectively adopting shop floor activities from production schedule. MES manages operation scheduling, equipment status management, production execution and control, shop floor supervision, material delivery and material consumption management. In order to accomplish manufacturing activities, MES need proper infrastructure for collecting data, analyzing data, integrating, and collaborating the manufacturing data. This is achieved in the operation level through supervisory control and data acquisition (SCADA) [4], a combination of hardware and software that monitors, gathers, and process the real-time data of the industrial processes within the organization or at remote place.

SCADA interacts with the smart devices and human-machine interface (HMI) software. Programmable logic controller (PLC) is used parallelly with SCADA. The PLC collects data from the input devices through sensors, process the data and initiates appropriate operations based on pre-defined settings. In Industry 4.0, PLC is used to communicate data through web browser, make connection with databases by structured query language (SQL) and link to cloud using message queuing telemetry transport. PLC and remote terminal units (RTU) helped in providing signals to the machines that are supervised and controlled by SCADA. Presently, IoT is transforming SCADA by contributing new standardization and openness. The characteristics of IoT platforms (scalability, interoperability, security) can be integrated in smart maintenance which results in improved efficiency, reduced waste, decreased downtime and extended equipment life. SCADA assists in human-machine interaction but, IoT is a machine-to-machine communication tool. Hence, IoT confirms that information is communicated to both machine and human to make every entity in the environment in aware of the scenario. The outcomes are economic benefit and less human intervention.

Industrial internet of things (IIoT) is a subcategory of IoT which specifically focus on industrial applications namely, manufacturing, oil and gas, agriculture and power plants. Gartner outlines IIoT as a collection of unified software capabilities. Using these capabilities, the industries improves their decision making in asset management, operational vista and control of plants and infrastructure [5]. The IIoT platform is designed to boost the security and safety related with the industry assets and its environment. Among the various important applications of IIoT in manufacturing, predictive monitoring and maintenance play a vital role. By applying machine learning (ML) algorithms, it can predict the failure in advance using the data collected from sensors. As the system predicts the time of failure, the order for the component that will fail in near future, will be placed and it is received at correct time by the maintenance team for replacing, to avoid machine downtime and improves overall efficiency. Another significant sector where IIoT can influence is Transportation and Logistics. The technical and monitoring system of IIoT improves the value chain system [6]. Thus, the industrial adaptation of IIoT will increase from 10% in 2020 to 50% by 2025 [7]. This chapter provides insights of IIoT in smart manufacturing. It includes the overview of IIoT-empowered smart manufacturing in Section 2.2. Section 2.3 discusses the various empowering technologies that made IIoT smart manufacturing a reality. Section 2.4 presents the framework for smart manufacturing and Section 2.5 converses about the case studies. Section 2.6 brings out the security issues and Section 2.7 sketches the challenges and opportunities in smart manufacturing. Finally, the chapter is concluded in Section 2.8.

2.2 IIOT EMPOWERED SMART MANUFACTURING

The IIoT is one of the vital technologies that add elegance to the industries. Smart factories and equipment with advanced intelligence built in are examples of the smart manufacturing sectors that use IIoT. Smart factories and equipment with advanced intelligence built in are examples of the smart manufacturing industries that use IIoT. Large-scale networks with heterogeneous devices connected to create dynamic states and generate enormous amounts of data are used in the smart manufacturing sector [8]. The networking system makes use of the internet, which adheres to open standards for communication and cloud technologies to securely access the devices. Big Data's sophisticated analytical tools are used for processing, while mobile technologies are used for quicker communication. Adoption of IIoT ensures improved security, efficiency, profitability, better safety, and performance with minimized carbon-di-oxide emission.

The IIoT Smart Manufacturing is characterized by pervasive digital connectivity that empowers the interaction among the systems, devices, machines, factories, and stakeholders. A general architecture of the IIoT system is discussed by Khan et al. [9] and it is shown in Figure 2.2.

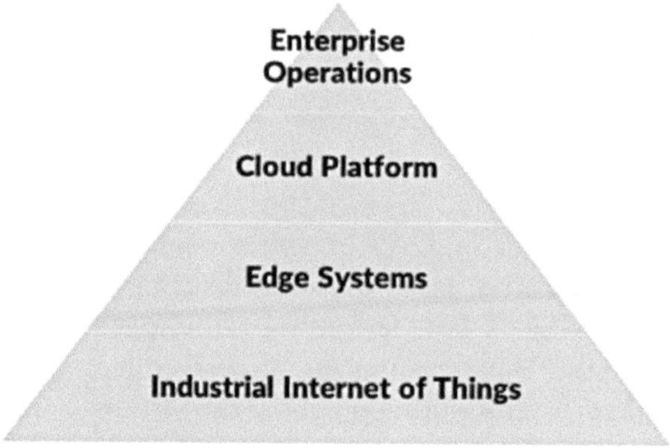

FIGURE 2.2 IIoT general architecture.

The IIoT General Architecture consists of four layers. The IIoT devices and the continuous stream of data generated by them are placed in layer 1. The connectivity management responsible for enabling communication between various devices are handled through the Edge systems in layer

2. The Cloud Platform in layer 3 is responsible for device authentication, storing data, performing data analytics, and decision making. The Enterprise applications, dashboards, APIs, and portals form the layer 4.

In the IIoT layer supports the communication between the smart devices deployed in this layer. The various IoT sensors such as accelerometer, gas, gyroscopes, humidity, Infra-Red sensors, motion sensors image, optical sensors, pressure, Radiofrequency Identifier, temperature are used to measure physical atmosphere and to acquire changes in the real-time environment. They are directly linked with the micro-processing unit, control unit, storage unit, wireless communication interfaces, and power system. These heterogeneous IoT devices have limited resources with different capabilities, therefore, different technologies like Wi-Fi, Wireless Sensor Network, Zig Bee, 4G/LTE/5G are employed for connectivity between smart devices. Power consumption is another important challenge that is solved through wireless recharging technologies during the critical events. Filtering, fusing sensed data using statistical, probabilistic, knowledge-based, inference, and reasoning methods [10] can be adopted to reduce the power consumption in transferring data to the Edge and Cloud layers. The IIoT device data is secured through registration with the Cloud Server. In order to access the device data, login operation is to be authorized by the Cloud Server and then the connection is established for further processing. Clusters are formed with cluster head to communicate with the Edge systems.

In a smart factory, the production devices generate massive data. Hence, the traditional Cloud data center find it difficult transmit and process the real-time data. In scenarios like emergency shutdown of equipment, predicament braking of smart connected vehicles in smart factory, the time-sensitive tasks that may lead to serious consequences, if the data processing is not performed with specific time frame. It is appropriate to apply Edge computing at these circumstances. The edge devices with adequate computing power in the Edge system execute native pre-processing of IIoT data to make immediate decisions and transmit the appropriate results to the Cloud center in milliseconds [11]. This improves the overall system performance by reducing system delay, demand for communication bandwidth and operational costs. Layer 2 allows the industries to install suitable security solutions locally to reduce the risk of data leakage during data transmission and data storage. So, the privacy and security threats are minimized.

The layer 3, Cloud platform of IIoT architecture is accountable for load balancing, device discovery, device identification and access management, routing, monitoring, and controlling of IoT devices. As Cloud demands payment only for the utilized resources, the cost needed in running an exclusive data center for the industry is reduced, which in turn increase the profit. Once the IIoT network is integrated with the Cloud, monitoring machinery in smart factories can be done from any location by the employees and managers. Problems in the automated assembly line is identified by the technician and notify immediately for equipment maintenance. This adds flexibility for the employees to work virtually from any location (off-site). One of the important properties of Cloud is scalability. This property enables the manufacturing operations to be performed at any scale using discrete manufacturing techniques. The discrete manufacturing technique empowers the manufacturers to construct products with low complexity in high volume and fabricate high complexity products in low volume. Thus, in smart manufacturing, scaling production up or down is carried out smoothly. Quality disputes can be solved on the factory floor, thereby reducing waste and modification cost. Microsoft offers Azure Solutions for smart manufacturing [12]. They state that data stored in Azure help the sales teams to improve and upturn their productivity, by knowing about their customer feedback and service data. Also, the sales team gains knowledge on sales trends and offer specialized scheme for the different categories of customers. So, Cloud assists in better decision making with respect to customer data. ML algorithms can be applied to the Cloud data to unveil the actionable insights related to product development, sales promotion, and customer satisfaction. Moreover, it may help the business to manufacture products based on customer demand and send the products to the customers earlier than the expected delivery.

The IIoT provides huge data for manufacturers. The information predicted from these data are to be presented to the user in user-friendly way through dashboards. There should be easy exchange of information between the user and machine. Data retrieval techniques and ML can be used for searching and getting real-time information from the IoT devices. This process can be automated through various APIs and ERP solutions. Intelligent ERP solutions, apply analytical tools on performance data of the machine and translate it into alters or warning to schedule the maintenance time for the machine during production. Operating the Industry in an efficient way needs to place right people at right location with precise information. Unified ERP solutions can be used for efficient organization of local teams with their job roles to get the work done effectively at right time [13]. The ERP solutions also assist to place, trace, and ship orders, manage, and store inventory globally. Though ML and ERP are two different mechanisms, the insights provided by the ML algorithms are employed by the ERP platform to communicate the prediction to corresponding departments and necessary actions are carried out at the right time. It also assists in decision making about the future activities.

2.3 ENABLING TECHNOLOGIES FOR IIOT

The strength of IIoT depends on different technologies such as IoT, edge cloud computing, big data analytics, blockchain, AI, CPS and augmented, and virtual reality (VR).

2.3.1 INTERNET OF THINGS (IOT)

In smart manufacturing, the real-time data collection is made through the IoT devices, and it forms the key element in IIoT. The IoT devices are deployed across warehouses, production, and distribution centers for monitoring and maintenance of the factory assets. The IoT devices follow the factory assets starting from raw material to finished products with the aim of reducing the labor cost and system management. Using IoT, a sustainable environment can be created. By the effective use of resources and connected IoT devices, the carbon footprint and pollution can be reduced.

2.3.2 CLOUD AND EDGE COMPUTING

The huge volume of data generated from IIoT demands extremely distributed high performance computing systems. Cloud computing provides compute, storage, and network services that can be adopted in an IIoT system for processing, storing, and analysis. Among the three Cloud service models viz., private, public, and hybrid, private cloud is implemented by the large enterprises to ensure privacy, safety and security and to cope with the competitive advantage. But small and medium sized enterprises may go for public Cloud deployment. In order to have real-time access, control, and monitoring of devices, IoT connects all smart things with internet. Edge computing is employed in IIoT to have interoperability between smart devices and to achieve access to the Cloud services by the smart devices [14]. Edge computing acts as an intermediate layer between IIoT and Cloud layer. It along with IoT gateway perform low level computations to reduce network traffic and response time. Additionally, it has facilities for fault tolerance and security.

2.3.3 BIG DATA ANALYTICS

SMSs produce a variety of data from interconnected smart devices. Traditional data processing techniques fail to process the stream of data produced by the IoT smart gadgets. Discovering new association and patterns in data, identifying persuading factors are possible on if big data analytics is applied [15]. Big Data empowers IIoT by:

- Generating knowledge for decision making from the continuous stream of real-time data created by the smart devices.
- Assisting in anomaly detection.
- Increasing the perceptibility of the smart manufacturing systems by earlier forecasts.
- Improving maintenance services and key performance indicator monitoring.
- Identifying customer demands and proposing new production cycle.

2.3.4 BLOCKCHAIN

The most threatening issue in IIoT smart manufacturing is the data security, device authentication and network isolation. Blockchain is a distributed ledger with potential for security, transparency, and fault tolerance. It is realistic to adopt Blockchain in IIoT systems to enhance the performance with respect to security and privacy. Several architectures were proposed by the researchers for trusted and resistive communication in IIoT to realize data assurance, flexibility, and accountability [16]. Blockchain is also used along with other technologies like anonymous encrypted messaging streams, multi-signatures to assure privacy and security. Supply chain issues in manufacturing such as ensuring the shipment process, precise forecasting of manufacturing materials, support for stocks and inventories, tracking sales through smart devices are handled safely and securely using blockchain [14]. Immutability characteristic of Blockchain assists in detecting malicious actions. Cloud services used by the IIoT manufacturing systems apply both public blockchain and private blockchain, at the service provider level, and at the workshop level, respectively, to collect data from every machine in the factory. The information that the manufacturers like to reveal to the floor operators can be controlled through the private blockchain.

2.3.5 ARTIFICIAL INTELLIGENCE (AI)

The IIoT system can be made to function as autonomous and intelligent system with the utilization of AI Technologies to improve efficiency and to reduce human-intercessions. Intelligence can be embedded in IIoT system at every layer beginning with sensors, edge servers and cloud data center by implementing various search, optimization, and prediction algorithms. Industrial robots can be employed to minimize the human-intercessions. Two important ML techniques, regression, and classification can be utilized to find out Remaining Useful Life and Time to Failure. These measures help to predict the machine failure in advance and thus the maintenance can be planned ahead. Isolation Forrest, K-means, Local Outlier Factor are the unsupervised models used for sensing anomalies in equipment [17]. The various ML and AI techniques can be applied to forecast the production depending on the trends in overtime and for quality control inspection. Thus, integrating AI and ML, enable the manufacturing system to customize the factory based on optimized operations and intelligent decision making.

2.3.6 AUGMENTED AND VIRTUAL REALITY (AR & VR)

Augmented reality (AR) and virtual reality (VR) technologies play a major role in IIoT smart manufacturing. It forms as an interface between stakeholders and the digital world of a smart factory. Instead of creating a physical prototype for a new product, VR is used to simulate the product. For example, Ford Motors designed an environment ford immersive vehicle environment (FIVE) [18]. Ford engineers use FIVE to design a car in virtual environment before its existence and test its performance virtually. VR and AR can simulate complex processes in industry operations. Hence, it is widely used in training the employees and provides directions to them during repair and maintenance. Though AI algorithms are used in the process of handling inventory databases, picking a product and organizing it in correct location within the warehouse is still a difficult task. A warehouse employee having an iPad or wearing a Microsoft HoloLens will be able to identify the place of a particular product and its exact location in the shelf by using AR technology. It reduces the search time in locating the product and for quick reply about the product to the user. VR can simulate the production and assembly line, to predict the threats that may cause disturbance. This assists in reducing the downtime, maintenance expenses and strengthen the employee security.

2.4 IIOT FRAMEWORK FOR SMART MANUFACTURING

As stated before, Smart manufacturing is a technology-driven approach where the production process in any industry can be automated, monitored, and controlled using Internet enabled machineries. Smart manufacturing concentrates in identifying the various industry operations that can be automated and makes use of data analytics to enhance the industry performance. Manufacturing is an explicit process in any industry and so, Smart manufacturing is a precise field of IIoT.

2.4.1 SIGNIFICANCE OF DATA IN SMART MANUFACTURING

Implementation of smart manufacturing can be initiated by embedding sensors in manufacturing machines. These sensors can be used to collect data about the machine's operational status and performance. The data collected are stored over a period of time and these data can be utilized by data analysts, manufactures, engineers, and data scientist in different ways. The data analyst can detect defects in particular machinery part by analyzing the data over time. If the analyst presumes a part may fail, then he can suggest the manufacturing engineers for replacement of that particular part that can avoid massive damage to the machinery and it can greatly reduce the downtime of the machine, which impacts in more production. The manufactures can use the data to analyze whether the current production setup is efficient or not. The manufacturers can clearly spot out where the production slows down or when the resources are under-utilized. Data scientists can utilize the data to run imitations of different processes to identify the well-organized ways of doing things.

When smart manufacturing is widely implemented, more machines have to be networked and the machines will be able to communicate with each other leading to potential level of automation. For instance, the smart manufacturing machines may order raw materials when there is not enough quantity available, or they may evenly distribute jobs among the machines. From this, we can easily interpret that data acts as a driving factor in smart manufacturing. In smart manufacturing, data decides "what has to be done" and "when can it be done." Smart manufacturing relies on data and its connectivity. Hence, securing these data also proves to be important. The IIoT framework for smart manufacturing has to focus on proper data management, data connectivity and data security.

2.4.2 DATA MANAGEMENT FRAMEWORK USING IIOT IN SMART MANUFACTURING

Compared with traditional manufacturing process, smart manufacturing is supposed to handle huge amount of data. Saqlain et al. [19] have projected a framework known as industrial data management system (IDMS) that collects real-time data, transmits the data, process the data and store the data is shown in Figure 2.3. As specified earlier, an efficient framework must give solutions for managing, transmitting, and securing data. The proposed framework, secured data management system for smart manufacturing (SDMSM) promises to provide the above. For convenience of understanding, the framework is divided into five basic layers where each layer possesses different functionalities.

The five layers are physical layer, secure communication layer, processing layer, storage layer, and application layer. The following sections will explain the functionalities of each layer in detail.

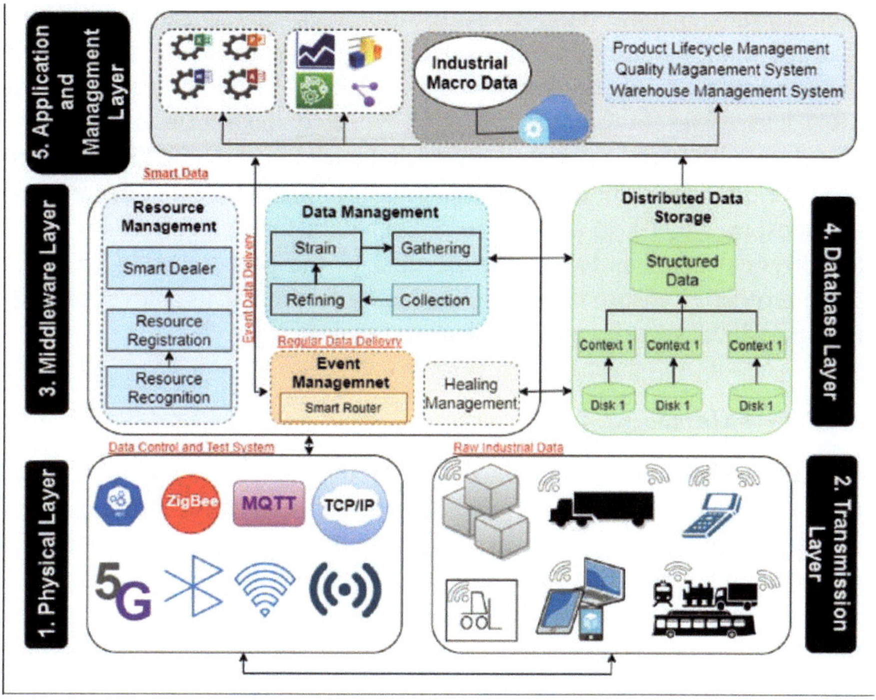

FIGURE 2.3 Secured data management system for smart manufacturing.

2.4.2.1 PHYSICAL LAYER

The physical layer is the layer where the real-time data are acquired by various entities like sensors, actuators, meters, robots, embedded chips, and various other devices. Once collected, they are communicated to the higher layer by the industrial communication protocol called open platform communication unified architecture (OPC-UA). Various forms of data are collected in real-time in digital format. Each device that transmits data are uniquely identified and any time the data can be acquired based on the device's identification.

2.4.2.2 SECURE COMMUNICATION LAYER

The industrial data that is collected by the physical layer is transmitted to upper layer with high throughput and low latency rate. Various communication technologies such as Bluetooth, RFID, Wi-Fi, ZigBee, etc., are used to transfer these data. These technologies have ability to store and forward small quantity of data over time. In order to assure data security and privacy, this layer deploys some protocols like IPv6, OPC-UA, SOAP, MQTT, etc. The IPv6 protocol is equipped to connect billions of devices at a time. OPC-UA is a standardized industrial communication protocol used for the consistent and safe transmission of fresh data collected from the sensors. MQTT is utilized to collect data from different devices in the physical layer and transmit the data to the processing layer. Using these protocols, ensure three levels of security such as user-level security, transport-level security and application-level security.

2.4.2.3 PROCESSING LAYER

The processing layer obtains the data from the physical layer and manages the data resources. This layer acts in the middle, there by communicating directly with the physical layer and the application/data storage layer. It drives all the requests and replies between these layers. Data management and data processing are performed at this layer. The various functions performed at this layer includes Data management, Event management, Resource management, and Recovery management.

Resource management component is responsible for data detection and arrange the data sources by the database fragment locations for storing, updating, and inquiring purposes. Data management is done by collecting the raw data and pre-processing it. After pre-processing, the data are aggregated and stored in the distributed databases. Once the data are collected from the smart devices, the event management decides whether the data has to be transported to the application layer (for concluding replies of crucial events) or for pre-processing and storing of data in the data management layer. Recovery management plays a vibrant role in resuming the data even if the database server goes down or crash suddenly. Data is stored periodically in remote servers, based on checkpoints. Checkpoints are nothing but fixed points and once this point is reached, the backup is taken until the checkpoint.

2.4.2.4 STORAGE LAYER

The aggregated data that comes from the processing layer is stored into local repositories in a distributed way. The main challenge lies in converting the heterogeneous industrial data into a structured data that can be easily stored. The data must be stored in a secured way in various distributed databases. Encryption techniques is applied before storing the data. That is, only unintelligible data is stored in the database. So, if anyone eavesdrops on the database, they couldn't get any valid data. Similarly, data is decrypted before it is utilized by the application layer. Data storage has to be managed properly in order to avoid duplication of data and loss of data.

2.4.2.5 APPLICATION LAYER

The application layer employs data analysis and data mining tools to extract meaningful patterns from manufacturing data and transform it to knowledge that can be utilized in the future. The extracted industrial data can be used in ERP, product lifecycle management (PLM), manufacturing execution system (MES), quality management system (QMS), supply chain management (SCM), and warehouse management system (WMS). Various deep learning (DL) and ML techniques can be applied on the industrial data for proper analysis.

2.5 CASE STUDY: IIOT IN SUPPLY CHAIN MANAGEMENT (SCM)

Supply chain management (SCM) is the management process of handling products or giving services to and for industries. The traditional supply chain process involves handling and supplying of raw materials, information, services, and processes. The supply chain can be done either within the industry or outside the industry. The conventional supply chain method has no balance between the stock keeping and customer requirement. In order to overcome the hurdles, SCM can be integrated with internet of things (IoT). This integration will benefit the industry in many ways:

- The information in SCM can be interconnected and communicated with smart objects.
- It is scalable. That is, both small- and large-scale industries can improve their performance.

- Most of the processes involved in SCM can be automated, e.g., inventory can be managed by using sensors that gives cent percent accuracy.
- SCM can be managed in real time using technologies such as RFID tags. Once these tags are scanned, information such as manufacturing date, expiry date, etc., can be deducted easily.
- Logistics-related information can be maintained with high transparency that eventually leads to consumer satisfaction.

In Ref. [20], a smart SCM (SSCM) framework is proposed that could satisfy the demands of any industry. SSCM uses RFID to track the products. SSCM framework shown in Figure 2.4, deals with Suppliers and Managers. The suppliers tend to provide all information related to the product and this information are stored in a database. Managers can access the database and know all the related information about the product. All enquiries related to the product can be sent to the supplier and the supplier can respond with necessary information. And again, all these communications can be stored in the electronic medium. The SSCM framework consists of three stages:

- **Stage I: Data Collection and Storage:** In this stage, a website is designed that acts as a means of communication between supplier and manager. RFID and other technologies can be used to track product information over time and this information can be stored in a readily available system database.
- **Stage II: Evaluation:** Here, the manager will evaluate the information stored in the database and select a product that matches his/her evaluation criteria. The criteria can be based on cost, quantity, quality, etc. After selecting the product, request can be done to the supplier and the supplier can respond by accepting or rejecting the request.
- **Stage III: Purchase:** In the final stage, upon accepting the product by the supplier, the manager can initiate the purchase process. The logistics operation can be made smart by equipping the transportation vehicles with global positioning system (GPS) to track the location of the vehicle and sensors to sense temperature, humidity, tire pressure, etc. This SSCM framework leads to smarter SCM than the traditional SCM.

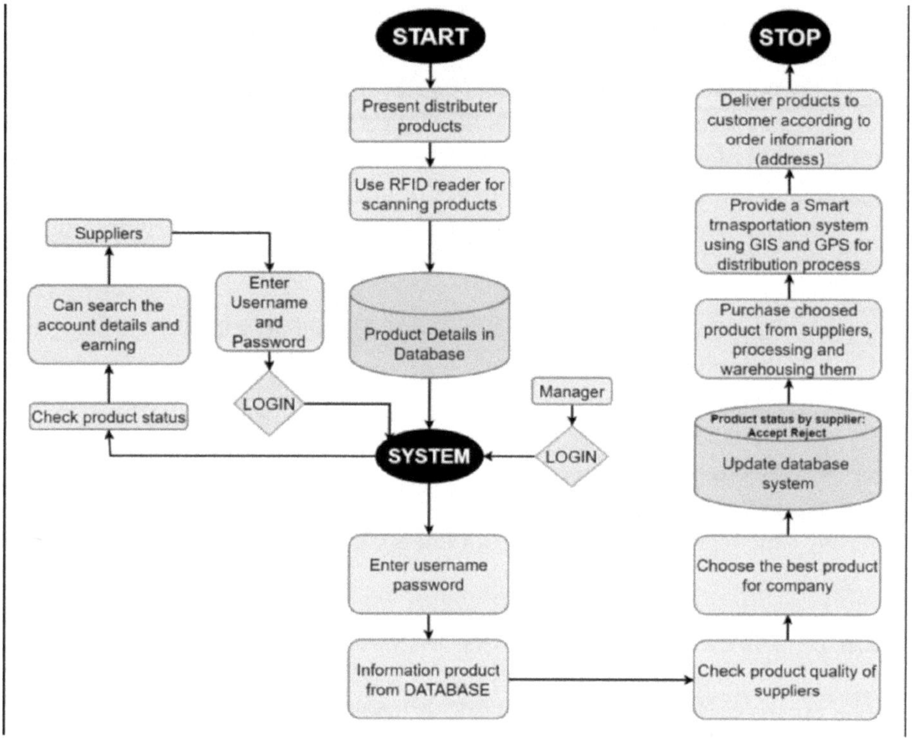

FIGURE 2.4 Smart supply chain management framework.

2.6 IIOT SECURITY IN SMART MANUFACTURING

Security ought to be preeminent in modern frameworks to deflect hurt and lessen the solicitation to staff, solid focuses, and the climate. There is extending industry recognizable proof that well-being and security are associated, for instance, that likeness a few realistic chances and plausibility and that destitution-stricken security threatens government assistance. This is likewise recognized in intercontinental viable security quality. With regular IACS frameworks, the utilization of the security standards depends on global principles, which support the utilization of a similar security point at defending top to bottom, particularly with the connection between various covering in the Purdue Model and isolation of tasks. The getting of IIoT erodes this fixed establishment by delivering current likeness from construction to undertaking or cloud-based frameworks, subsequently becoming the forthcoming for well-being and security-associated breaks.

A new shortfall of viable ways to deal with the consolidation assessment of well-being and security hazards is fundamental in ordering IIoT results [21]. A composite structure has been recommending, however further work is needed to systematize its utilization and test its immaterialness in mechanical plants. Information security is opening up countless associated gadgets to the cloud offers various section focuses for an assailant.

Security Act of 2007, there are fundamental characteristics in the security act [22]:

- Responsibility liberal by hold out against attacks;
- Prunella vulgaris volume;
- Active development;
- Upgrade accuracy;
- Strength of features;
- Federation of demand feedback;
- Incorporation of giving out energy sources.

2.6.1 SAFETY POINT IN SMART MANUFACTURING

The creative assembling framework utilizes an incorporated organization framework in an assembling framework to divide data among assembling or assemble units to the termination clients. For this reason, it needs network availability and is organized exceptionally through the web. Sharing intelligence data through the web includes information safety and data through the framework at different focuses with extraordinary worldwide recognizable proof and start-to-finish information encryption. Furthermore, thus, every organization's hub ought to be secured against outer assaults and information abuse. In particular, while planning arranged frameworks, for example, canny assembling frameworks guarantees the security of the framework [23] and the general interaction.

2.6.2 NETWORK PROTECTION

The approach of the IIoT is speeding up the requirement for network protection in modern control frameworks. The intricacy of IIoT will imply that network protection should be planned into the segments that make up the mechanization framework. The appropriation of mechanical security guidelines with affirmation will be crucial for the progression of IIoT. It will guarantee the security of individual resources and the more powerful frameworks and frameworks of frameworks. These confirmations will assume a part like those which happen in the domain of security accreditations. Observance to certification implies that the components of a framework hold the critical security building blocks. The fixings are consolidated safely by security-ensured groups and are worked as a safe framework by security-prepared administrators. Augmented administrators are more valuable because they get the correct data at the perfect time. Obstructions to appropriation. The selection of mechanical security principles with accreditation is fundamental for the headway of IIoT.

2.6.3 THE INDUSTRIAL INTERNET OF THINGS (IIOT)

An advancement to an elegant Connected endeavor Schneider voltaic. The way to security confirmation is consistency and pertinence. Around the world, the IEC62443 arrangement of safety norms covers all security components from item advancement to item includes framework highlights, conveyance, and activity. It is observed that even IEC has not embraced any of these bodies for IEC62443 accreditation.

In the first place, the information safety and security of the IIoT gadget has been given by utilizing upgraded RSA and hash marks. The following commitment is the bunching method for the sensor gadgets utilizing the NDRF-SSA procedure dependent on sensor hub degree, distance from the group head, lingering energy, and wellness. The third commitment is to devise grouping using SoftMax-DNN for best asset booking dependent on capacity, figuring, and transfer speed prerequisite.

2.7 IIOT CHALLENGES AND OPPORTUNITIES IN MANUFACTURING

2.7.1 CHALLENGES

The industrialists and engineers face numerous challenges in implementing the new technologies. Those issues are explained below [24]:

1. **Security Issues:** The primary challenge is Security issue on IIoT technologies. Dr. Irene Petrick, Director of manufacturing transformation, Intel's IoT classification, specified that fast growth

in inventive technologies like connected machines, globalized computer technologies, smart devices, robots, and real-time data analytics in the more extensive manufacturing system expect from IIoT require to report possible provocation and the security concerns in implementation. The proscription of complete cyber-security solutions is the main concern of large-scale manufacturing systems. The leading issue is that the abuses will affect both individuals and company in financial and operational activities. As manufacturing sectors use various IIoT suspension, they are unlocked to safety cover, cyber risks related with information technology (IT), active technology convergence, and insider threats. Appling IIoT automation will be confronted till it has a strong secure characteristic.

2. **IIoT Integration:** Operational technology (OT) and integrating IT is a critical challenge encountered by organizations during Industrial IoT application. Almost attention should be taken while incorporating these technologies and confirm that the data communication is made without any loss and susceptibility. IIoT incorporation requires a tremendous effort for its application which may be unpractical for many working companies. Typically, IoT appliance are expended as free-spirited result, and it is adopted into the engineering and fabricate procedure to become a portion of the whole structure. However, combination of OT and IT lacks effective relatedness and harmonization.

3. **Connectivity and Visibility:** Another complex challenge during the application of IIoT is the connectivity problems between different components and the management of whole system. Regular problems are faced in integration, due to power loss, internet interruption, physical, and scientific mistakes. This will isolate the linked devices from the digital network, the entire production process may be disturbed leading to an extra cost in damages. It is essential to check connected machines and tools in real-time, and engineers should certify that those machines/tools are functioning with optimum manufacture ability. Observing the lives of the equipment, understanding the problem and solving the issues before they occur is also critical.

As the framework of fabricating structure suits elegant, they are incrementally functioning and implemented by a growing number of devices. We perceive that different device may execute the like and unlike purpose or tasks. Some devices depend firmly on the production of additional appliances, just similar to a channel outcome line. Such powerful link may also differ an energetic, turning on the non-identical assignment implementation. In a word, the alliance of several devices has set off censorious for the all-inclusive production of existing and time to come structure. The internet of manufacturing things (IoMT) arrange a host of infrared to monitor the device state uninterrupted. Sensor result, known as machine stamp, supply an unrivaled chance for excellent settlement making in fabricating. However, perceiving the brimming prospective of IIoT for smart manufacturing be dependent on a considerable expanse on super scribing the provocation. The provocation is the position of each device. This position and being engaged or not and good physical condition, in the perception of whether it is purpose accurately or not. This estimation detail is crucial as it regulates whether these devices can be added up for job implementation.

The most straightforward strategy is to utilize sensors to detect and investigate the detected information signal. A wired stock or batteries may control these sensors. Notwithstanding, with the expanding ordinal number of devices, thinking about their contemplated lifespan of 1 to 20 years, it is trying to give dithery force or battery hold up in certain situations. For instance, wires restriction the compactness of sensors. Battery substitution is additionally, in some cases, testing and tedious for these detecting frameworks. Further, batteries may not be protected or productive in some limited conditions. The subsequent test is how to utilize the situation with different machines to convey errands to every gadget. Should each device follow a stern unchanged timetable? The likely glitch of machines and robust exchange in framework extent positions will bring about programs contrastingly streamlined towards energy, absolute activity time, and so forth. Under these conditions, conveying undertakings powerfully to machines dependent on the detected status of every gadget is fundamental. The third test starts from the interchanges between devices, conceivably including the organizing machines on the off chance they exist. It is noticed that these days a few undertakings are done by devices from various origin, even from different nations. This uncovers the opportunity of issues in correspondence dependability and its effect on the coordinated effort in the middle of these devices. Also, fabricating resources are shut frameworks that can't be controlled altogether from an external perspective, regardless of whether a two-path data stream exists. Take a mechanical device, for instance. One can dispatch G code to the device to execute it, yet one can't straight forwardly authority to the servers and shafts of the gadgets. This is one more colossal test that should be defeated to empower complete oversight and mechanization. Addressing these difficulties will prompt new roads of crucial and applied examination in Manufacturing, manageable Manufacturing, Cognitive Supply Chains, and IIoT.

2.7.2 OPPORTUNITY OF IIOT [25]

- **Opportunity 1:** Retrofit Heritage Devices for Imaginative Mass Produce: Although new assembling advances and upstart organizations emerge, many existing assembling firms fall behind the flood of computerized development. It isn't phenomenal that heritage devices exist in numerous tiny assembling ventures. While heritage machines are essential resources for assembling experience and are entirely used underway, they need continuous and in-measure detecting and control frameworks. Accordingly, little producers progressively lose the upper hand in the worldwide market since they are restricted in data permeability and the capacity to adapt to the more remarkable intricacy of present-day fabricating conditions. Significant to the issue is building up IoT availability between heritage machines. IoMT detecting gives an extraordinary chance to retrofit inheritance machines for advanced Manufacturing. Thus, there is an immediate need to foster new attachment and play IoT sensors that ceaselessly gather in-site machine information, send the information to distributed depository, and speak with other "things" and partners.
- **Opportunity 2:** Self-Controlled Machine Status Detecting: The IIoT retrofitting of inheritance devices for intelligent manufacturing. It is essential to perceive the nature of a gadget with the identity-controlled stock instrument. By production, the detector self-controlled, no radio association or accumulator is expected to give a force contribution. With extra remote matching and information change, such a detecting framework could be utilized

proficiently in numerous devices, improving compactness and decreasing upkeep fetch. Wave preparing procedures, further essential crude information handling or start-to-finish execution of capacities, could likewise be attach to the sensor hub.

- **Opportunity 3:** Device Administration and Assignments Planning and Circulating: There is a chance to examine upgraded task appropriation (booking) system between a gathering of machines for many undertakings or administrations considering the help of detected machine information. For instance, in a robust framework with a worker community, there will include the accompanying inquiries to be replied to:

  - When and what errands ought to be conveyed to which machines (in a unique dissemination framework)?
  - How to survey the expected commitment of a machine that is failing yet might be fixed? What will this mean for task planning and appropriation arrangements?
  - What are the vitality utilization and usage of each device?

- **Opportunity 4:** The Collective Energy in the Middle of IoMT Machines: It is likewise fundamental to upgrade the collective power betwixt many gadgets working together distantly. Because of actual space and untrustworthy communication channels, there might be a brief square in correspondence in the middle of the machines in better places. How should a device respond when it ends up disengaged from different gadgets? How could the focal organizer be planned?

- **Opportunity 5:** Distributed Enumerate and Investigation: The specter information stage is a unified information vault, including authentic information gathered from numerous machines and online news from the gadgets in-situ. This information can be recovered effectively from the cloud stage to neighborhood PCs to separate valuable data and model calculations sent in further the cloud or the IoT sensor gadgets [26]. Information operate framework information processing and authority is an imperative advance to the up-and-coming age of computerized Manufacturing. Distributed computing and examination will open roads of freedom to advance the administration and arrange assembly tasks, from quality administration, power the executives, warmth and cooling, maintainability and security, appropriation, and store network the board.

- **Opportunity 6:** Blockchain-Empowered IoT: IoMT things speak with one another through the web. Information reliability and protection arise as a significant matter for the plan, improvement, and IoT frameworks. The assembling business is a piece of the basic framework of every country. Digital assaults on the IoMT framework will straightforwardly disturb producing tasks and other fundamental capacities in appropriate areas.

2.7.3 DIFFICULTIES FOR INDUSTRY 4.0

It isn't challenging to accomplish Industry 4.0, and it is probably requiring at least 10 years to work out. This is a future vision, since it includes several viewpoints, innumerable troubles and complications, including logical problems, creative challenges, financial challenges, social issues, and policy-driven issues [27]. This chapter discusses a portion of the science and innovation challenges, like creating clever gadgets, building the organization climate, broad information investigation and preparing, and advanced creation. A brief illustration of these troubles is specified below.

Various threats of Industry 4.0 [28]:

1. **The Improvement of Clever Gadgets:** Manufacturing plant working with Industry 4.0 advancements is intensive and will require more intervention from counterfeit machines (counting robots), lessening the inclusion of representatives. Be that as it may, various plants need distinctive creative gadget designs, and savvy gadget advancement requires a lot of time and cash before being placed into creation in an Industry 4.0 manufacturing plant.
2. The development of the organizational climate in Industry 4.0 essentially introduced in the CPPS stage utilizing CPS innovation. Building the CPPS stage is an unpredictable scheme restricted by different circumstances [29].
3. **Huge Information Investigation and Preparation:** All of Industry 4.0, including the gear, machines, applications, items, and administrations, will produce huge and complex information.

Proficient constant mix and information examination in the current creation cycle should be guaranteed to streamline assets in the creation chain. In the meantime, it is necessary to address how huge the information offerings risk in data insurance and protection issues. Examination and applications in sizeable mechanical details are not incredibly experienced, and all regions, from apparatus to software design, need additional enhancement.

4. Advanced assembling development ought to amplify digitization, re-enactment, and Internet advances to realize equipment joining, software design reconciliation, and a merged data set. Numerous innovations, the panel, principles, and different challenges will be experienced during the implementation interaction. As the electronic realm and the actual realm become incorporated, network security issues are getting increasingly extreme for Digital Manufacturing. There are abundant issues concerning network security and digitalization can be accomplished to forestall computer scientists and different cybercrimes brought about by the encroachment.

2.8 CONCLUSION

The IIoT system permits the manufacturing industry to gather and inspect a huge data to improve the overall administration of the systems. This chapter provides the insights of IIoT in smart manufacturing. It includes the overview of IIoT empowered smart manufacturing and discussed the various empowering technologies that made IIoT smart manufacturing a reality. The framework for smart manufacturing for secured data management is conversed and the implementation of IIoT in SCM was thrashed out. The challenges faced by Engineers and Industrialists such as security issues, IIoT integration, connectivity, and visibility are sketched in this chapter. These challenges can be oppressed by recommending proper solutions. This chapter can be considered as a recommendation to address some of the uncertain challenges in the field of IIoT through recent technologies like Edge, Blockchain, and Cloud.

KEYWORDS

- **blockchain**
- **data management framework**
- **industrial internet of things**
- **internet of manufacturing things**
- **security act**
- **smart manufacturing**
- **supply chain management**

REFERENCES

1. Monostori, L., Botond, K., Thomas, B., Shinsuke, K., Soundar, K., Gunther, R., Olaf, S., et al., (2016). Cyber-physical systems in manufacturing. *CIRP Annals, 65*(2), 621–641.
2. *Industry 4.0: The Fourth Industrial Revolution – Guide to Industry 4.0.* https://www.i-scoop.eu/industry-4-0 (accessed on 2 November 2023).
3. Ghahramani, M., Yan, Q., MengChu, Z., Adrian, O. H., & James, S., (2020). AI-based modeling and data-driven evaluation for smart manufacturing processes. *IEEE/CAA Journal of Automatica Sinica, 7*(4), 1026–1037.
4. *Internet of Things and SCADA: Is One Going To Replace The Other?* https://iiot-world.com/industrial-iot/connected-industry/internet-of-things-and-scada-is-one-going-to-replace-the-other/ (accessed on 2 November 2023).
5. *Industrial IoT Platforms Reviews and Ratings.* https://www.gartner.com/reviews/market/industrial-iot-platforms (accessed on 2 November 2023).
6. Karmakar, A., Naiwrita, D., Tapadyuti, B., Manojeet, C., & Md, R., (2019). Industrial internet of things: A review. In: *2019 International Conference on Opto-Electronics and Applied Optics (Optronix)* (pp. 1–6). IEEE.
7. Eric, G., Scot, K., Ted, F., Alfonso, V., Emil, B., & Amitesh, S., (2019). *Magic Quadrant for Industrial IoT Platforms,* ID G00363999, Gartner.
8. Conway, J., (2016). *The Industrial Internet of Things: An Evolution to a Smart Manufacturing Enterprise*. Schneider Electric.
9. Khan, W. Z., Rehman, M. H., Hussein, M. Z., Muhammad, K. A., Nasrullah, A., & Khaled, S., (2020). Industrial internet of things: Recent advances, enabling technologies and open challenges. *Computers & Electrical Engineering, 81*, 106522.
10. Krishnamurthi, R., Adarsh, K., Dhanalekshmi, G., Anand, N., & Basit, Q., (2020). An overview of IoT sensor data processing, fusion, and analysis techniques. *Sensors, 20*(21), 6076.

11. Qiu, T., Jiancheng, C., Xiaobo, Z., Zhaolong, N., Mohammed, A., & Dapeng, O. W., (2020). Edge computing in industrial internet of things: Architecture, advances, and challenges. *IEEE Communications Surveys & Tutorials, 22*(4), 2462–2488.
12. *How to Optimize Smart Manufacturing Production with IIoT and the Cloud.* https://cloudcheckr.com/cloud-automation/smart-manufacturing-iiot-cloud/ (accessed on 2 November 2023).
13. Sean, B. *How to Leverage the IIoT and ERP to Create a More Connected Enterprise.* https://www.encompass-inc.com/how-to-leverage-the-iiot-and-erp-to-create-a-more-connected-enterprise/ (accessed on 2 November 2023).
14. Younan, M., Essam, H. H., Mohamed, E., & Abdelmgeid, A. A., (2020). Challenges and recommended technologies for the industrial internet of things: A comprehensive review. *Measurement, 151*, 107198.
15. Nagorny, K., Lima-Monteiro, P., Jose, B., & Armando, W. C., (2017). Big data analysis in smart manufacturing: A review. *International Journal of Communications, Network and System Sciences, 10*(3), 31–58.
16. Liang, X., Juan, Z., Sachin, S., & Danyi, L., (2017). Towards data assurance and resilience in IoT using blockchain. In: *MILCOM 2017–2017 IEEE Military Communications Conference (MILCOM)* (pp. 261–266). IEEE.
17. Zheng, H., Antonio, R. P., & Chris, S. G., (2020). *Advancing from Predictive Maintenance to Intelligent Maintenance with AI and IIoT.* arXiv preprint arXiv:2009.00351.
18. Nitin, T. *Bridging the Digital and Physical Worlds Using AR/VR and Industrial IoT.* https://www.pratititech.com/blog/bridging-the-digital-and-physical-worlds-using-ar-vr-and-industrial-iot (accessed on 2 November 2023).
19. Saqlain, M., Piao, M., Shim, Y., & Lee, J. Y., (2019). Framework of an IoT-based industrial data management for smart manufacturing. *Journal of Sensor and Actuator Networks*, *8*(2), p.25.
20. Abdel-Basset, M., Manogaran, G., & Mohamed, M., (2018). Internet of things (IoT) and its impact on supply chain: A framework for building smart, secure, and efficient systems. *Future Generation Computer Systems, 86*, 614–628.
21. Abuhasel, K. A., & Khan, M. A., (2020). A secure industrial internet of things (IIoT) framework for resource management in smart manufacturing. *IEEE Access, 8*, 117354–117364.
22. Phuyal, S., Bista, D., & Bista, R., (2020). Challenges, opportunities, and future directions of smart manufacturing: A state of art review. *Sustainable Futures, 2*, 100023.
23. Boyes, H., Hallaq, B., Cunningham, J., & Watson, T., (2018). The industrial internet of things (IIoT): An analysis framework. *Computers in Industry, 101*, 1–12.
24. Kumar, A. S., & Iyer, E., (2019). An industrial IoT in engineering and manufacturing industries–benefits and challenges. *International Journal of Mechanical and Production Engineering Research and Development (IJMPERD), 9*(2), 151–160.
25. Yang, H., Kumara, S., Bukkapatnam, S. T., & Tsung, F., (2019). The internet of things for smart manufacturing: A review. *IISE Transactions, 51*(11), 1190–1216.
26. Badarinath, R., & Prabhu, V. V., (2017). Advances in internet of things (IoT) in manufacturing. *IFIP International Conference on Advances in Production Management Systems (*pp. 111–118). Springer, Cham.

27. Thoben, K. D., Wiesner, S., & Wuest, T., (2017). Industrie 4.0 and smart manufacturing-a review of research issues and application examples. *International Journal of Automation Technology*, *11*(1), 4–16.
28. Zhou, K., Liu, T., & Zhou, L., (2015). Industry 4.0: Towards future industrial opportunities and challenges. In: *12th International Conference on Fuzzy Systems and Knowledge Discovery (FSKD)* (pp. 2147–2152). IEEE.
29. He, T., (2014). Industrial 4.0 strategy: The challenges and prospects of CPS. *Mobile Communications,* 19–21.

CHAPTER 3

Learning Using Neural Networks for Smart Mobility

E. P. SUMESH,[1] S. R. RAMYA,[2] R. VIDHYALAVANYA,[3] and V. YATHAVARAJ[4]

[1]*Professor, School of Electronics Engineering (SENSE), VIT-AP University, Andhra Pradesh, India*

[2]*Department of Computer Science and Engineering, Dr. NGP Institute of Technology, Coimbatore, Tamil Nadu, India*

[3]*Department of Computing and Electronics Engineering, Middle East College, Muscat, Oman*

[4]*Assistant Professor, Department of Computer Science and Engineering, Dr. NGP Institute of Technology, Coimbatore, Tamil Nadu, India*

ABSTRACT

Modern lifestyles demand smart mobility to reduce time, effective use of resources, and cost. Most of the time the data is processed manually to manage the mobility in rail, road, and other means of mobility. Recent development in technology made huge amount of data available in various form such as text, video, images, etc. Processing of huge amount of data is practically impossible manually, and this demands artificial intelligence (AI) to arrive at decision making to improve mobility aspects. Recent development in neural network and its ability to make decisions accurately, has opened up many applications including smart mobility.

Smart Mobility and Intelligent Transportation Systems for Commercial and Hazardous Vehicles.
T. Poongodi, Naga Swetha Pasupuleti, and Balamurugan Balusamy (Eds.)

In this chapter, a brief review of neural networks (NNs) and its application in smart mobility is presented. Internet of things (IoT) with deep learning (DL) NNs has revolutionized mobility concepts in terms of safety, transport flow, human mobility, parking, and choice of transport network in smart cities. A detailed description is also provided on application and design of NNs. Choice analysis is another promising DL NNs application. Text data analysis is to infer traffic disruption in rail system is another important application of neural network in automating mobility. Further combination of spatio-temporal analysis with neural network also will be discussed as an improvised method.

3.1 INTRODUCTION

Emergence of IoT resulted in smart solutions covering a wide variety of applications backed up by technological development in communication, embedded systems and big data analytics. Availability of wide varieties of sensors at affordable cost has resulted in deployment of IoT systems in smart home, smart logistics and transportation, smart healthcare systems, smart power and smart city management, etc. [1, 6]. Massive amount of data available through IoT devices increases the challenges to interpret the data and to make appropriate decisions leading to efficient system implementations. Conventional machine learning (ML) technology cannot be used in such a scenario due to enormous data processing and reduced performance. Deep learning (DL) can be effectively applied to overcome these stumbling blocks raised due to massive data; rather DL takes the advantage of availability of high data to make wise decisions to manage the system [2].

3.2 DEEP LEARNING (DL) NETWORK FUNDAMENTALS

Deep neural network (DNN) consists of numerous layers having competent learning structures based on input data. Each layer of DL network consists of processing units called neurons. Structure of a fundamental neuron is given in Figure 3.1. Neuron output is the weighted sum of all the inputs and is then given as input to an activation function which produce the desired output. In Figure 3.1, x_1,x_2, ... x_n represents the inputs to the neuron. The weights w_1, w_2, ... w_n are the weights used to scale the inputs.

The variable *b* represents the bias which can be considered as an offset value.

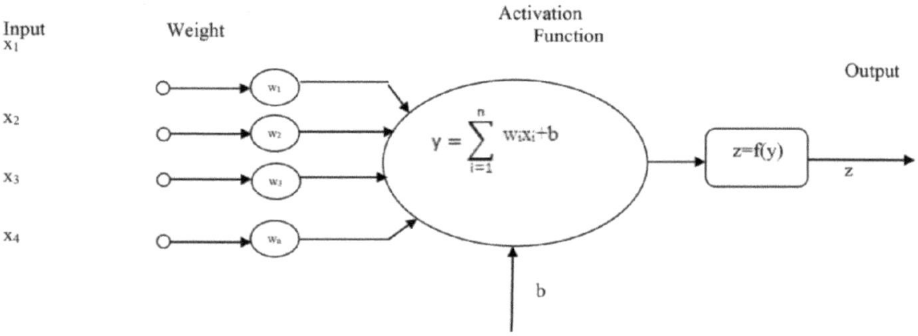

FIGURE 3.1 Artificial neuron.

Neuron output is then given as input through an activation function *z*. Some of the functions used as activation function is provided in Table 3.1.

3.2.1 DEEP LEARNING (DL) NEURAL NETWORK ARCHITECTURES

Several DL network architectures are available to fit into real world problems. Some of the notable architectures are:

- Convolutional neural networks;
- Deep belief networks;
- Stacked de-noising auto-encoders;
- Feed-forward neural networks;
- Recurrent neural network;
- Recursive neural networks (RNNs);
- Multi-layer perceptron (MLP);
- Self-organizing maps;
- Convolutional deep belief networks;
- Deep Boltzmann machines (DBM).

An overview of DNN is given in Figure 3.2.

TABLE 3.1 Few Activations Functions

<table>
<tr><th>Sl. No.</th><th>Activation Function</th><th>Graph</th></tr>
<tr><td>1.</td><td>Identity</td><td>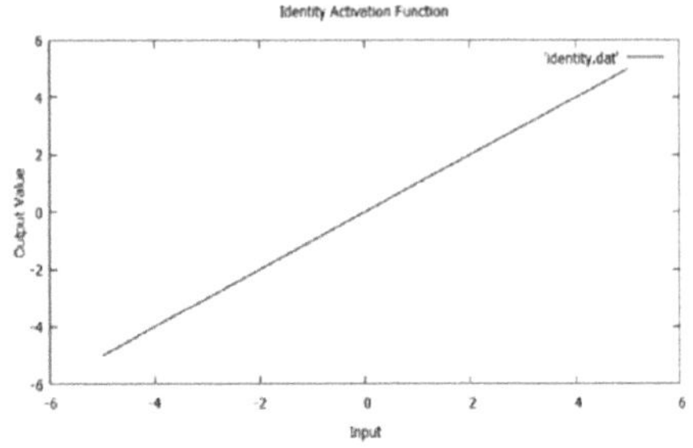
</td></tr>
<tr><td>2.</td><td>Hyperbolic tangent</td><td>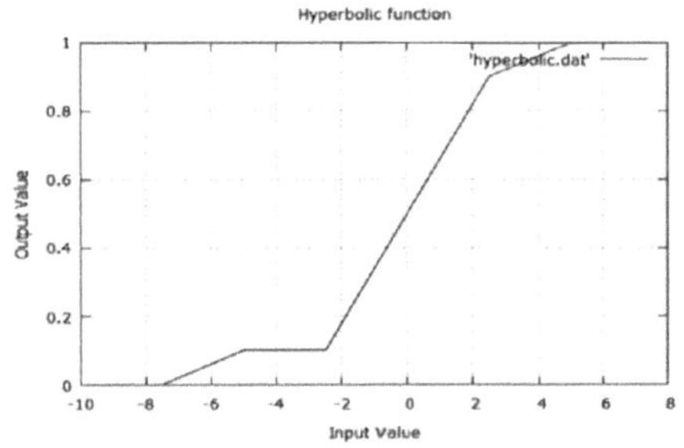
</td></tr>
<tr><td>3.</td><td>Sigmoid</td><td>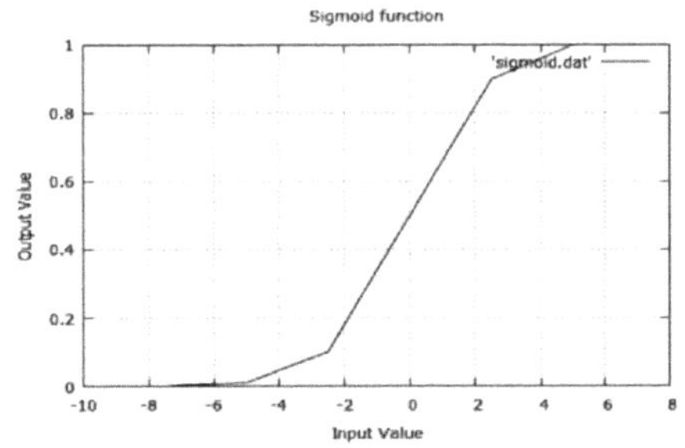
</td></tr>
<tr><td>4.</td><td>Rectifier linear unit (ReLU)</td><td>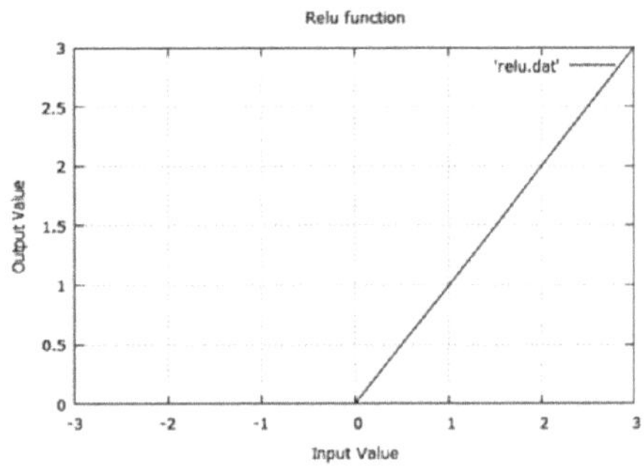
</td></tr>
</table>

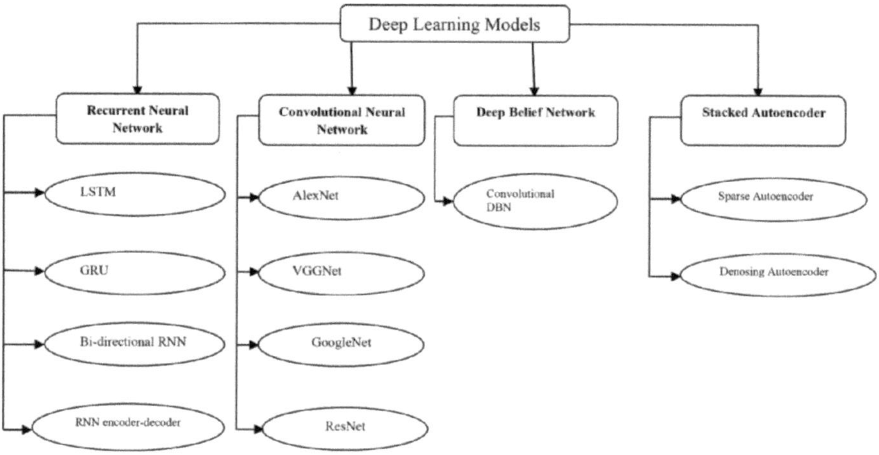

FIGURE 3.2 Overview of deep learning network.

3.3 LITERATURE SURVEY

Many traditional methods were presented by researchers for the transportation system. AI plays a vital role for transportation and is advantageous than the traditional methods. It can automatically predict the traffic conditions and using several sensors. In Refs. [39–41], various AI methods were proposed used for transportation using ANN like road planning, public transport, etc. Supervised and Unsupervised learning methods were used in predictions of the future traffic [42–49]. Optimization problem is the major issue reported which needs to be solved to make the analytics easier. Advanced computational algorithms like genetic algorithm (GA) [50–53] and simulated annealing [50, 54–56] were used to solve the optimization issues. Few AI methods were developed with the concept of working as groups as defined by bees and ants. ACO [57, 58] and BCO [59–62] methods were also helpful for solving the complex optimization issues in transport. Shortest path optimization is a needed concept for an easy and speed means of transport. In Ref. [63], it was proved that fuzzy logic model (FLM) outperformed logistic regression model (LRM). For a perfect prediction and adaptability these algorithms are helpful for dynamic traffic situation. It was reported in Ref. [64] that for the multiple cases and scenarios, agent-based software engineering (ABSE) is useful. In several research papers, the tools used for AI for transport and planning of travel automatically was explained. These methods will be helpful for the people to improve the travel plan without any congestion and improve reliability.

Understanding the road traffic and finalizing the correct method as the need of the society is important in the current transport system. So, a design plan method is first approach before solving the complex issues. Network design problem (NDP) [65] is a suitable method for designing an optimal road. The oldest methods used were NNs for planning, designing, and modeling. But these methods require nodes and links. Due to these disadvantages, new raster algorithms are used [66]. Due to heavy traffic in the urban area, ML algorithms are developed. GA and SA algorithms were used for high and low-demand situations, respectively. In Ref. [53], it was reported that GA outperforms SA by having less computational time.

Designing of the transport system means the preparation of the framework of transport management with safety plan and balance the demand of the highways. The authors compared several algorithms like GA and SA [49] and demonstrated that GA outperforms SA.

3.4 CONVOLUTIONAL NEURAL NETWORK (CNN)

Convolutional neural networks (CNNs) are a feed-forward type architecture which is widely used for image processing applications. It contains different layers of processing. Figure 3.2 shows a typical CNN architecture to identify vehicle types.

It consists of several layers of convolutional layers followed by suitable activation functions. To reduce the special size of convolved feature, pooling layers are used. This step helps to reduce the processing load since the input signal being an image consists large number of values after the convolution step. It also serves the purpose of extracting dominant features effectively in the training phase. Two kinds of pooling step are frequently used. Max pooling is used to extract the maximum values from the portion of images operated by the convolutional kernel. This step also suppresses noise components to certain extent. Another method is average pooling which reduced the dimensionality by extracting the average value from the image portions operated by the kernel. After going through several stages of convolution and pooling, depending on the complexity of the network, output data are flattened and fed to conventional neural network for final categorization.

A classic network scheme is shown in Figure 3.3 to demonstrate the overall flow and dimensionality reduction.

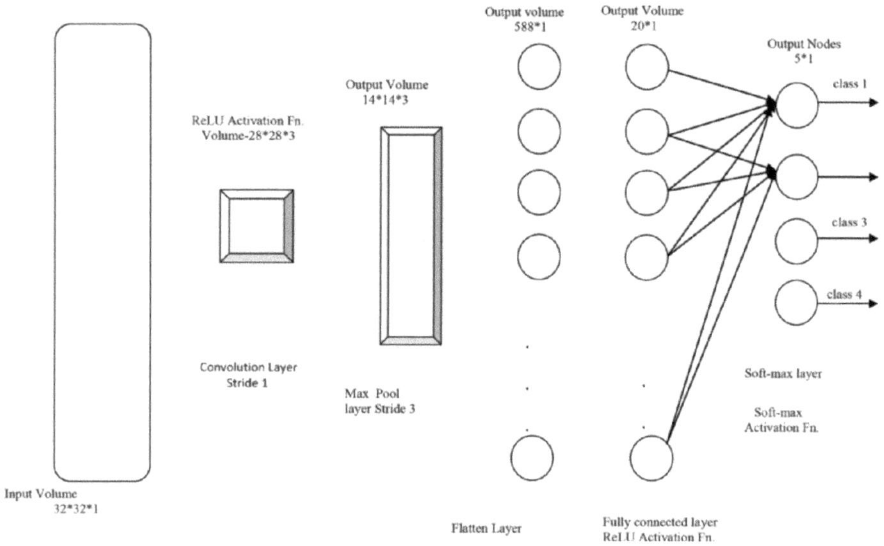

FIGURE 3.3 Illustration of CNN architecture and dimensionality reduction.

In Figure 3.3, a grayscale input image of size 32×32×1 is given to the convolution layer with stride 1. The generated image which is a 28×28×1 is passed through a max pool layer which takes average of every 2×2 boxes thus produces a 14×14×1=588 values. This output is flattened to a one-dimensional array of size 588×1. Output from the flatten layer is connected to a fully connected (FC) activation layers and mapped to the desired output classes [3]. FC layers are used at the output side since the important features are captured by previous layers. Some of the commonly used CNN architectures are: LeNet, AlexNet, VGGNet, GoogLeNet, ResNet, and ZFNet.

3.5 RECURSIVE NEURAL NETWORKS (RNNS)

In recursive neural network (RNN) same set of weights are recursively used to process a structured input. This process happens in each parent nodes and children are nodes similar to the parent node as illustrated in

Figure 3.4. Due to this property, RNN is widely used for predicting structured output [4, 5]. Some of the applications includes natural language processing (NLP) and sentiment analysis.

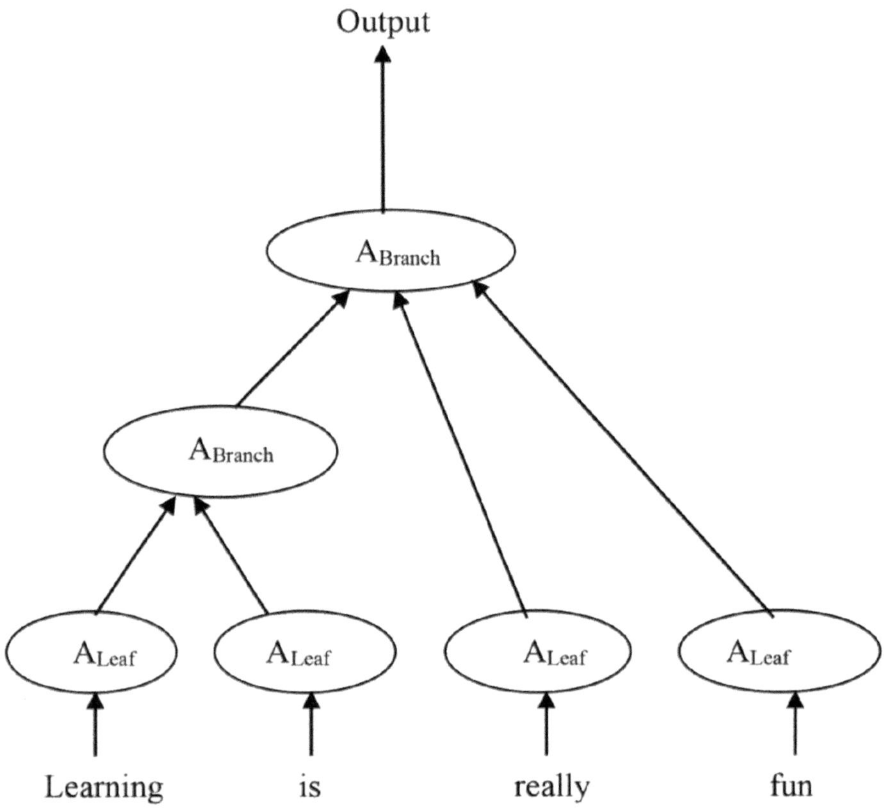

FIGURE 3.4 Recursive neural network structure.

3.6 DEEP BOLTZMANN MACHINES (DBM)

Deep Boltzmann machines (DBM) consists of several layers of hidden variables. Data dependent expectations estimation is done with variational approximations. General structure of a DBM is given in Figure 3.5.

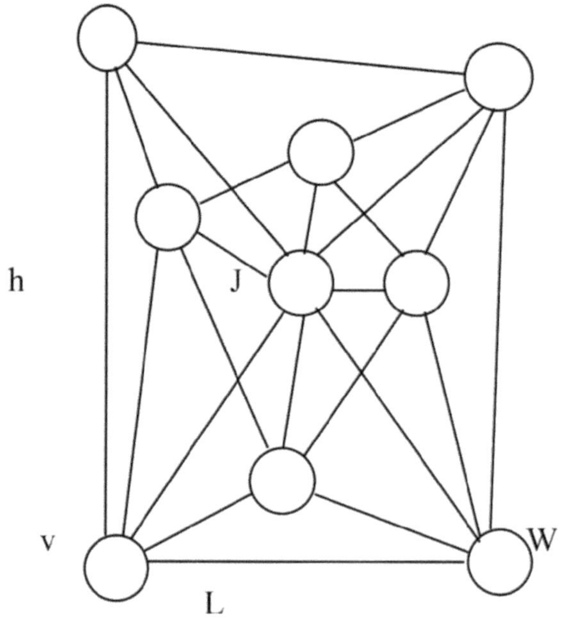

FIGURE 3.5 Structure of a DBM system.

Some of the salient features of DBM systems are:

- It contain visible units and multiple hidden layer units;
- It has uni-directional connection or no intra layer connections;
- Layer by layer pre-training is possible;
- From each layer back propagation is used to fine tuning.

3.7 APPLICATION OF CNN IN INTERNET OF VEHICLES (IOV)

Internet of vehicles (IoV) using optical camera communication (OCC) is a promising application to establish IoT like scheme for smart transportation [9–11]. The aim of IoV is to improve the traffic safety with efficient systems which links vehicles to wireless router and/or mobile nodes to create a large network and process the data to derive the objectives of smart transportation system.

Overview of IoV system is given in Figure 3.6.

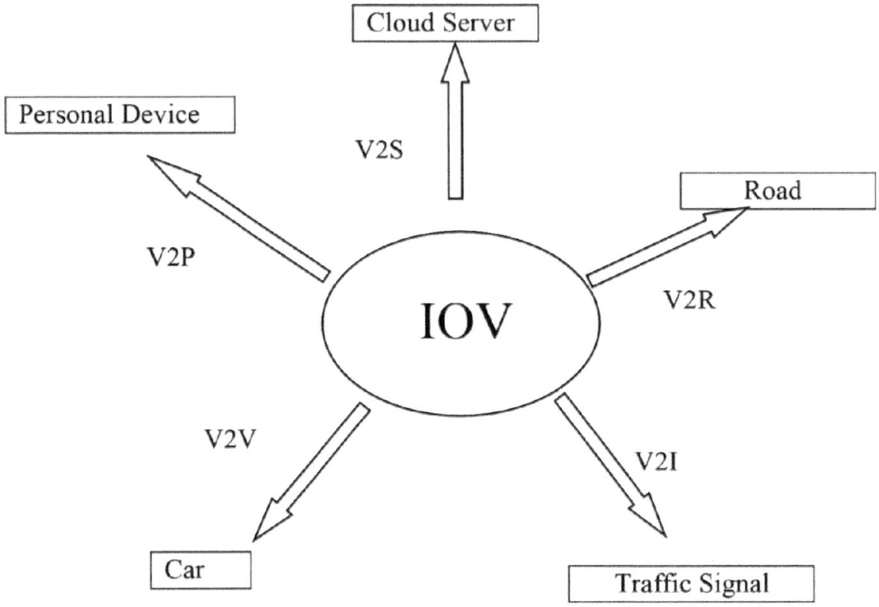

FIGURE 3.6 IoV system components.

IoV connects various components using IoT. The heterogeneous architecture consists of vehicle-to-vehicle (V2V), vehicle-to-cloud (V2C), vehicle-to-roadside unit (V2R), vehicle-to-infrastructure (V2I), vehicle-to-sensors (V2S), and vehicle-to-pedestrians (V2P). Entire system banks on information exchange with intelligence [12]. LED signals are used as transmitter whereas a camera serves purpose of a receiver.

One of the objectives of IoV is achieved by collision avoidance and thereby reducing accidents. To achieve this information pertaining to surrounding infrastructure and vehicles with high precision and reliable communication is essential.

Optical wireless communication (OWC) employing image sensors and LEDs is considered as a promising and cost-effective solution to establish IoV [9]. Some of the key requirements in IoV include road-sign detection, blind-spot detection, backup cameras, road side LED detection, distance measurement from other vehicles and infrastructure, lane departure monitoring, detection of obstacles, etc.

The success of the system depends heavily on development of efficient hardware and intelligent processing of data and making safe decisions.

Development of hardware is a separate research area and efficient decision-making system is focused here subsequently.

The images acquired by the camera are prepossessed by image processing algorithms to extract the region of interest. The distance parameters are obtained using stereo vision cameras [14] employing two cameras to capture a stereo image pair.

3.8 CNN-BASED IMAGE PROCESSING

DNNs plays an instrumental role in implementation of IoV – specifically recognition of objects based on features captured by camera or a similar sensor. Figure 3.7 shows steps used in processing of data.

In step 1, preprocessing of data is done to remove noise from captured image and necessary scaling and conversions. The preprocessed data is the passed through the CNN to select ROI, this step detects objects and do necessary filtering, feature extraction and image segmentation. Further refinement of ROI done in next step by recognition, tracking, feature mapping and finally pattern recognition. ROI and Refinement of ROI is done through a combination of CNN and image processing steps. Finally, a decision making is done to analyze and initiate necessary control actions.

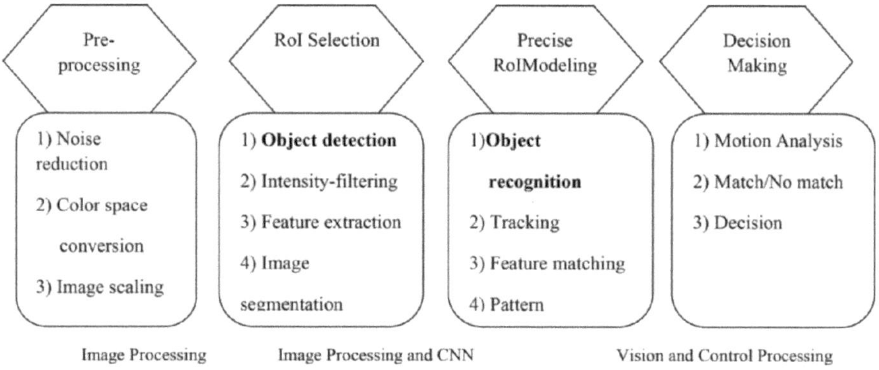

FIGURE 3.7 Object recognition using CNN.

Figure 3.8 describes the CNN architecture. The camera produces a 32×32 image, after image preprocessing mentioned earlier input images are fed as input to Layer 1 of CNN. The middle layer in the network consists convolutional filters, ReLU, and pooling layers. During the training phase weights of convolutional layers are refined according to the information of the image. Pooling layers reduces the feature size, and the number of hidden layers can be adjusted to an optimum and careful down sampling by pooling layers to obtain desired performance measures are the challenges in designing the network. After a number of hidden layers FC layers are employed towards the output stages. The network can be trained to detect emergency conditions and broadcast them through the cloud to warn other vehicles to take measures for the emergency.

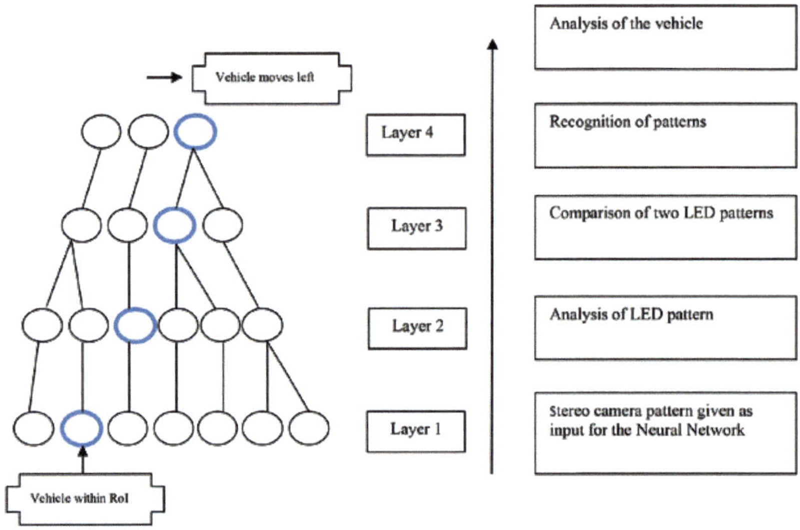

FIGURE 3.8 CNN architecture for the vehicle approach classification and recognition.

3.9 TRAFFIC FLOW FORECAST BASED ON DEEP LEARNING (DL) TECHNIQUE

Traffic forecasting is an area where lot of research has been taking place, intelligent transportation with DL and IoT is one among the promising technology in recent times. Prediction and analysis of traffic volume by historical average, K-nearest neighbors algorithm (KNN), time series, ARIMA and Kalman filtering method are some of the techniques employed.

Fast graph convolution recurrent neural network (FastGCRNN) provides a promising solution to traffic flow forecast. FastGCRNN model for traffic prediction is given in Figure 3.9. In this model traffic flow in each node is taken as input sequence 'X' capture by sensors [14] whereas the flow of traffic in each node in the future is represented as output 'Y.' The spatial structure information about the road through convolution followed by sampling to reduce dimensionality is extracted by FastGCN. FastGCN unit is followed by the gated recurrent unit (GRU) which captures the long- or short-term temporal dependence between X. Encoder unit consisting of GRU and the hidden layer. Decoder predicts the traffic flow at each node continuously [13]. In Ref. [13], an experimental study on 1,865 roads in Luohu District of Shenzhen city was presented and information to obtain data set is also provided.

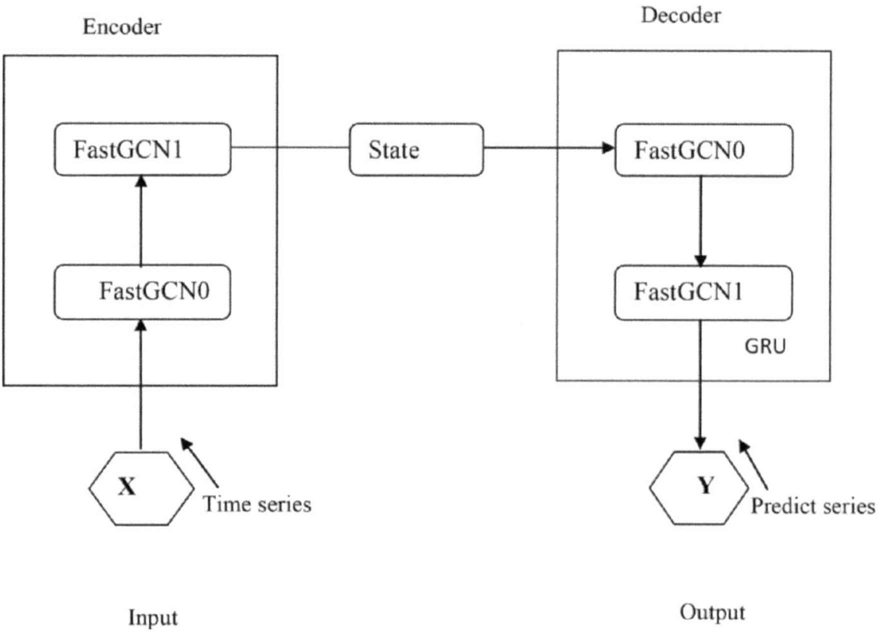

FIGURE 3.9 Fast graph revolution recurrent neural network model for traffic congestion prediction.

3.10 DNN IN CHOICE ANALYSIS

Owing to the high prediction ability of DNN it plays an important role in smart mobility applications like predicting travel mode choice, automobile ownership, fastest service, travel scheduling, travel frequency, queue length, route choice, etc. [15–18].

3.11 DNN MODEL FOR CHOICE ANALYSIS

A typical CNN model for choice analysis is given in Figure 3.10. It consists of an input and output layer and 7 hidden layers with 100 neurons in each of the hidden layers [15].

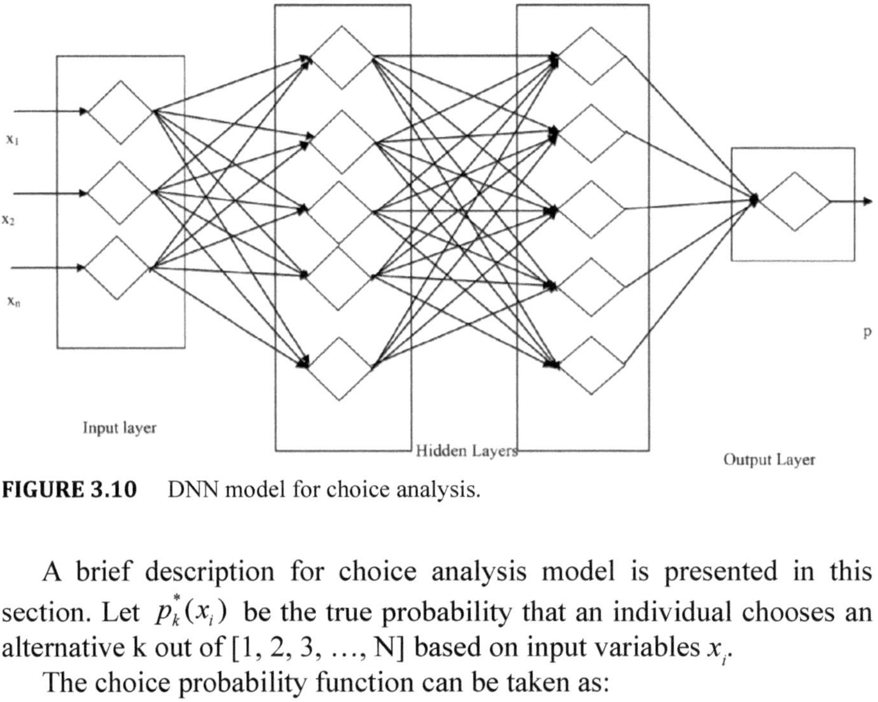

FIGURE 3.10 DNN model for choice analysis.

A brief description for choice analysis model is presented in this section. Let $p_k^*(x_i)$ be the true probability that an individual chooses an alternative k out of [1, 2, 3, …, N] based on input variables x_i.

The choice probability function can be taken as:

$$p_k(x_i) = \frac{e^{V_{ik}}}{\sum_j e^{V_{ij}}}$$

where; V_{ik} and V_{ij} are the jth and kth input values of the Softmax activation function.

where; $V_{ik} - g_m{}^k \circ gm_{-1} \circ g_{m-2} \ldots \circ g_2 \circ g_1(x_i)$ and $g_i(x) = \mathrm{Re}LU(w_l x + b_l)$

where; w_l and b_l and are the weights and bias of the network, respectively as given in Figure 3.1.

Though DNN can be treated as a universal approximator, it has large estimation error owing to the high approximation power [15]. Hyper parameter tuning in terms of depth, width, and model complexity need to be effectively to arrive at an acceptable estimation error.

3.12 DEEP LEARNING (DL) IN TRAFFIC DISRUPTION MANAGEMENT A CASE STUDY

Traffic disruption data is normally processed manually through texts, documents or emails. This type of data is unstructured one and mostly filled into predetermined templates. Automation of these data by artificial intelligence (AI) will benefit the agencies to identify trends, faults in procedures, customer demands, scheduling, introducing new business strategies and operations, etc.

A typical case study on automation using historical data is presented in Ref. [20] and how to extract delays, impacts on trains, potential actions to prevent unexpected incidents and classification of incidents into predefined classes are discussed. Technique used in Ref. [20] is described briefly in this section.

- **Step 1: Word Embedding:** In this step dense representation of words to capture semantic and syntactic information is carried out. bidirectional encoder representations from transformers (BERT) performs the embedding effectively. BERT can be done either by a feature-based approach or by a fine-tuning based approach.

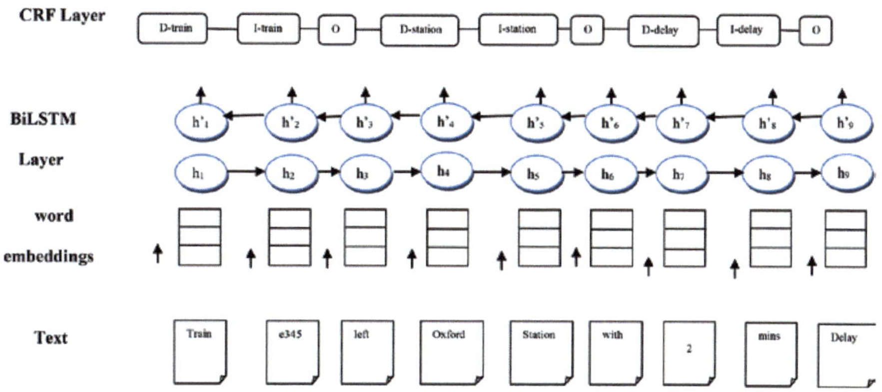

FIGURE 3.11 BiLSTM CRF method for NER.

In BiLSTM CRF model each input text x is assigned to a tag. Let $x = \{x_1, x_2, x_3, ..., x_n\}$ *and* $y = \{y_1, y_2, y_3, ..., y_n\}$ represents the input tokens and their corresponding tags.

BiLSTM computes two hidden values h_t and h_t of the sentence extracting the left and right perspective at each word. Then h_t and h_t are concatenated at the CRF to make an effective representation of the word. After the BiLSTM layer, a linear layer predicts the score $e_t = \tanh(Wh_t)$

To capture dependencies between consecutive tags CRF is used. If P is the score matrix for a tag and $T_{i,j}$ indicates the score transitioning from the tag i to j. The score of a sentence can be evaluated as:

$$s(x, y) = \sum_{i=0}^{n} T_{Y_i, Y_j} + \sum_{i=0}^{n} P_{i, Y_j}$$

The training is done to maximize the log probability [21]:

$$\log(p(y \mid X)) = s(X, y) - \log\left(\sum_{\bar{y} \in Y_X} e^{s(X, \bar{y})} \right)$$

where; Y_X represents set of all tags.

3.13 TOPIC MODELING

Step 2: Topic Modeling: In this step organizing and understanding of textual documents are carried out with unsupervised learning algorithm. Probabilistic Latent Semantic (pLSI) [22] and Latent Dirichlet Allocation (LDA) [23] are two types of algorithms that can be used for topic modeling. Consider a document with n distinct words $W_d = \{W_{d,1}, W_{d,2}, W_{d,3}, ..., W_{d,n}\}$. If there is K number of topics, each $k \in K$ has a distribution β_k over the words in the dictionary. A high β_k indicates the word has high probability of occurrence under a given topic. β_k is a multinomial distribution with Dirichlet prior parameter η. The document d is generated with a division of topics in K with a multinomial parameter θ_d and a parameter α.

Algorithm for LDA is described below:

- Make a distribution over words B_k~*Dirichlet*(η) for each topic k.
- For a document d:

- Construct a vector of topic proportions $\theta_d \sim Dirichlet(\alpha)$.
- For a word $W_{d,n}$ make a topic $Z_{d,n} \sim Multinomial(\theta_d)$, $Z_{d,n} \sim Multinomial(\theta_d)$, and a word $p(W_{d,n} \mid Z_{d,n}, \beta)$. Here, $p(W_{d,n} \mid Z_{d,n}, \beta)$ is a multinomial probability distribution on topic Z_n
  where; η and α are the hyper parameters to tune the network [24].
- **Step 3: Document Classification:** In this step document is automatically labeled when a dataset containing document *d* and label *c* is received; this step need to be a supervised one.

In Ref. [20], an application based on Transport for London (TfL) data base containing incidents over a time period was developed. In their study the used TextBlob to preprocess the data like spelling correction, conversion to lowercase, removal of punctuation, etc. Further stop words and numbers were removed and data was converted to tokens (sentences) and different forms of same word is converted to a common word like – canceled, cancel, and cancellation are represented by same word cancel.

With a 75% and 25% split for training and testing BERT model was used for word embedding. LSTM model with 32 hidden units were employed for the model. F1 score provides the model accuracy and a F1 score between 0.88 and 0.94 was reported in Ref. [20] and incorrect manual reporting was rectified by the algorithm.

DNN structure for the incident classification.

Figure 3.12 describes the DNN architecture for incident classification. The text data is converted to numerical features and fed to DL network. In Ref. [20], it is reported that the Pre-trained BERT embeddings considerably increased the logistic regression and deep learning (resnet) models' prediction accuracy for analysis of textual data.

3.14 MULTITASK LEARNING DEEP NEURAL NETWORKS (MTLDNN) FOR CHOICE ANALYSIS

Multitask learning deep neural networks (MTLDNN) can be effectively used for a combined analysis of revealed preference (RP) and stated preference (SP). Figure 3.13 shows the architecture of MTLDNN. This architecture has three shared and task-specific layers each, it has M1 shared and M2 task-specific layers. In Task 1 it uses RP whereas Sp is used in Task 2.

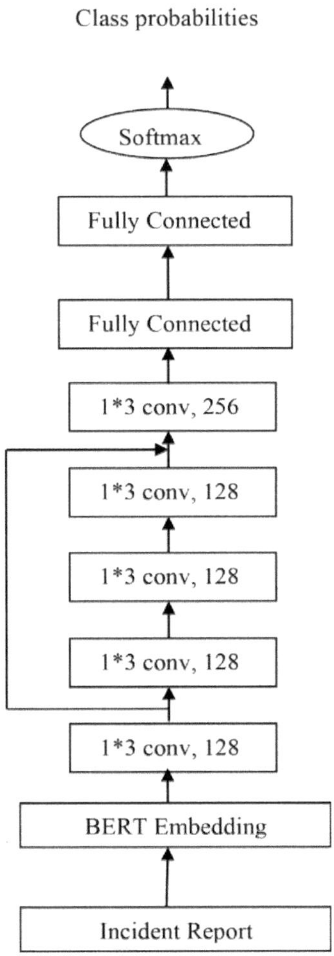

FIGURE 3.12 DNN architecture for text classification.

In Ref. [25], choice analysis based on autonomous vehicles (AVs) data from Singapore was implemented. The data was derived from a survey consisting of three parts RP, SP, and respondent's demographics. In RP, zip code of origin and target of recent trip with a reason along with travel choices. The time of travel was estimated by Google Map API.

In SP, hypothetical choice scenarios was included. The respondent's age, gender, and income are included in the demographics. MTLDNN with hyper parameter tuning was employed to maximize the prediction

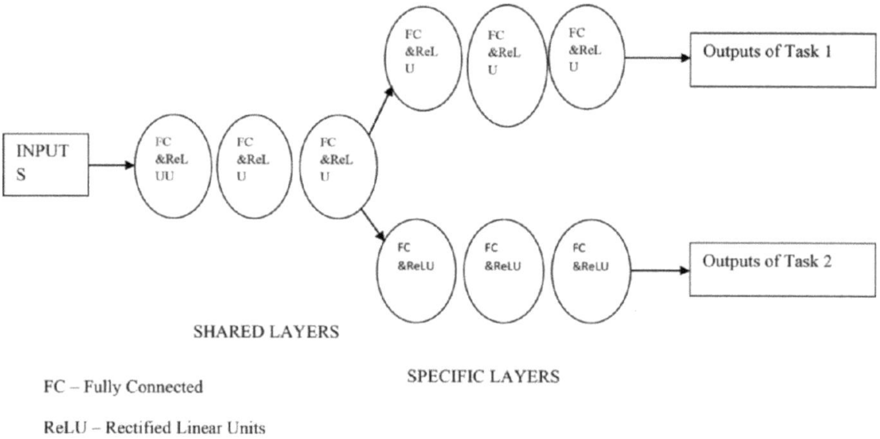

FIGURE 3.13 MTLDNN architecture.

accuracy. MTLDNN achieved an accuracy between 60.3% and 70.7% as per experiments done in Ref. [25], however it has lower performance in terms of cross entropy losses. There is a scope of further research to develop a new DNN architecture to perform consistently over various performance measures as well as to improve the accuracy.

3.15 HUMAN MOBILITY PATTERN DETECTION USING DNN

Presently the number of person using mobile device crossed 3.6 billion in 2020 as per the survey conducted by satista.com [36]. Along with advancement in number of mobile users' rapid development in IoT resulted in huge amount of data generated by several millions of machines nodes. Along with mobile users their usage in social media, watching videos, online shopping, etc., also has increased.

For a data scientist this data is a gold mine for developing multitudes of applications. As mobile devices are often carried by people, it can be used for estimating the mobility pattern of users. Study of mobility pattern is very important in many situations like epidemic control, transportation, and facility management and planning.

CNN can be effectively used to analyze and predict human mobility pattern due to its local dependency properties. As there is a strong dependency between nearby pixels in an image as nearby positions in the human mobility are often relative. Another aspect for suitability of CNN is its scale invariance; images in human trajectory might have different scales. CNN has the ability to preserve spatial features invariant with respect to temporal phases.

Various studies reveals that personal mobility is not random, mobility of human mostly obeys the reproduction mode. Ouyang et al. [37] developed a deep-space model to predict human mobility where the authors attempt to find where human goes next? For achieving the objective optimal feature set was derived from the data set by merging similar features thereby leading to an optimal feature representation and prevention of overfitting. The CNN model-deep-space [37] was divided into sub-models for input and output sections as shown in Figure 3.14.

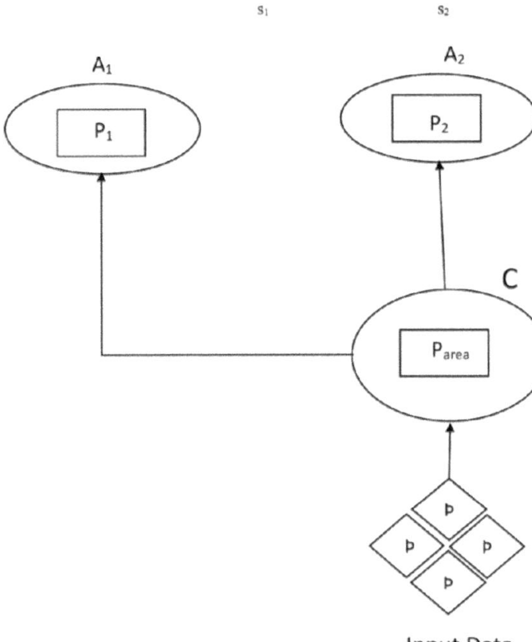

FIGURE 3.14 Deep space architecture.

In Figure 3.14, the sub-model D is the course prediction model and the sub-models $\{S_1,S_2,S_3...,S_n\}$ are the fine prediction models. Mobile base station longitude and latitude are used for exact position coordinates. Each base station was given a prediction label $\{L_1, L_2, L_3 ..., L_n\}$.

When the mobile data *M* has arrived, it is passed through the model *D* which is a CNN, the coarse prediction model locate the person in a crude range of position and labeled as L_{area}. Subsequently, the data is processed in next higher sub-model *S*. Let us consider new data label L_{area} and past data label L'_{area} are different then it is possible to run the algorithm parallel.

In Ref. [37], an experiment using data derived from a south-eastern city of China – containing usage detail records (UDRs) of mobile communication – was presented. The data has 40 fields describing the human activities in the network. Based on the base station locations the longitude and latitude are represented by x-coordinate and y-coordinate represents, respectively. A data cleaning step was performed by removing the missing data and removal of duplicate base station switching details and the cleaned data fed to the CNN architecture shown in Figure 3.15.

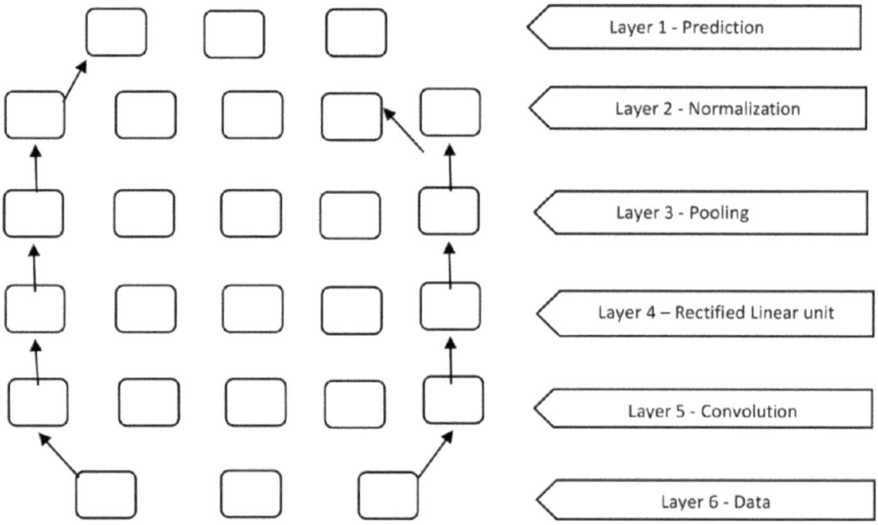

FIGURE 3.15 Deep space CNN architecture.

The CNN architecture has 4 hidden layers containing convolution layers, parametric rectified linear units (PReLU), max pooling layers and normalization layers.

In Ref. [37], the model was run for 50-to-200-time intervals with a step size of 50-time intervals and achieved an overall accuracy of 70 to 71% approximately 15% better than conventional methods.

3.16 FUTURE SCOPE

An overview of suitability of DNN in different mobility application is provided in Table 3.2.

TABLE 3.2 Summary of DNN and Applications

Model	Input Type	Applications
RNN	1D	Traffic flow, speed, and human activity [26].
CNN	2D	Human mobility or activity, parking slot images, vehicle images, disaster images [27–30]
DBN	1D	Traffic flow [31, 32]
SAE	1D	Traffic flow [33, 34]
CNN+RNN	2D or 3D	Vehicle trajectories [35]

Smart city applications are of promising research areas to be focused on future. Combination of various networks with parametric optimization can be considered as a future direction in smart city application. Choice analysis with higher accuracy and consistence performance with respect to various parameters is an open challenge for researchers. Another area to be explored are feature engineering, techniques like wavelet transforms can be used to obtain better feature sets than feeding the data directly to the DNN structure [38].

3.17 CONCLUSION

In this chapter, the way how the decision making can be improved using AI have been discussed and added to that recent development in neural network and techniques to make decision accurately have been discussed. The application of smart mobility has been discussed. Major concern in the concepts like safety, transport flow, human mobility, parking in smart city has been revolutionized using IoT with DL have been discussed. Furthermore, application and design of neural network have been discussed in detail. Another important application like choice analysis, analysis of text in rail system in neural network for automating mobility has been discussed. Finally headed over the combination of spatio-temporal analysis with neural network have been discussed.

KEYWORDS

- **artificial intelligence**
- **internet of things**
- **mobility**
- **neural network**
- **secure communication layer**
- **smart manufacturing**
- **supply chain management**

REFERENCES

1. Ejaz, A., Ibrar, Y., Ibrahim, A. T. H., Imran, K., Abdelmuttlib, I. A. A., Muhammad, I., & Athanasios, V. V., (2017) The role of big data analytics in internet of things. *Computer Networks, 129,* 459–471.
2. Panetta, K., (2016). *Gartner's Top 10 Strategic Technology Trends for 2017.* [Online]. Available: http://www.gartner.com/smarterwithgartner/gartners-top-10-technology-trends-2017 (accessed on 2 November 2023).
3. Sumit, S., (2018). *A Comprehensive Guide to Convolutional Neural Networks — the ELI5 Way.* [Online]. Available: https://towardsdatascience.com/a-comprehensive-guide-to-convolutional-neural-networks-the-eli5-way-3bd2b1164a53 (accessed on 2 November 2023).
4. Alejandro, C., (2009). *Understanding the Principles of Recursive Neural Networks: A Generative Approach to Tackle Model Complexity.* [online] available: https://arxiv.org/ftp/arxiv/papers/0911/0911.3298.pdf (accessed on 2 November 2023).
5. Hammer, B., Micheli, A., & Sperduti, A., (2007). Adaptive contextual processing of structured data by recursive neural networks: A survey of computational properties. In: Hammer, B., & Hitzler, P., (eds.), *Perspectives of Neural-Symbolic Integration.* Springer.
6. Navamani, T. M., (2019). *Deep Learning and Parallel Computing Environment for Bioengineering Systems* (pp. 123–137). Academic Press.
7. Salakhutdinov, R., & Geoffrey, H., (2009). Deep Boltzmann machines. *Proceedings of the Twelfth International Conference on Artificial Intelligence and Statistics,* [Online]. Available: http://www.utstat.toronto.edu/~rsalakhu/papers/dbm.pdf (accessed on 2 November 2023).
8. Salakhutdinov, R., & Larochelle, H., (2010). Efficient learning of deep Boltzmann machines. *Journal of Machine Learning Research — Proceedings Track, 9,* 693–700.

9. Islam, A., (2018). Convolutional neural network scheme–based optical camera communication system for intelligent internet of vehicles. *International Journal of Distributed Sensor Networks, 14*(4).
10. Cun-Qun, F., Shang-Guang, W., Zhe, G. W., et al., (2013). Enhanced throughput multipath routing algorithm based on network coding in IoVs. *J. Commun., 34*(Z1), 133–141.
11. Yang, F., Wang, S., Li, J., et al., (2014). An overview of internet of vehicles. *China Commun., 11*(10), 1–15.
12. Benrhaiem, W., Hafid, A. S., & Sahu, P. K., (2016). Multi-hop reliability for broadcast-based VANET in city environments. In: *Proceedings of the 2016 IEEE International Conference on Communications (ICC)*. Kuala Lumpur, Malaysia. New York: IEEE.
13. Zhang, Y., Mingming, L., & Haifeng, L., (2020). Urban traffic flow forecast based on FastGCRNN. *Journal of Advanced Transportation, 2020*, 9. Article ID: 8859538. https://doi.org/10.1155/2020/8859538.
14. Mauri, A., Khemmar, R., Decoux, B., Ragot, N., Rossi, R., Trabelsi, R., Boutteau, R., Ertaud, J. Y., & Savatier, X., (2020). Deep learning for real-time 3D multi-object detection, localization, and tracking: Application to smart mobility. *Sensors, 20*(3), 532.
15. Peter, N., Aura, R., & Tommaso, T., (1996). Modeling inter-urban transport flows in Italy: A comparison between neural network analysis and logit analysis. In: *Transportation Research Part C: Emerging Technologies* (Vol. 4, No. 6, pp. 323–338).
16. Subba, R. P. V., et al., (1998). Another insight into artificial neural networks through behavioral analysis of access mode choice. In: *Computers, Environment, and Urban Systems* (Vol. 22, No. 5, pp. 485–496).
17. Giulio, E. C., & Stefano De, L., (2005). Multilayer feedforward networks for transportation mode choice analysis: An analysis and a comparison with random utility models. In: *Transportation Research Part C: Emerging Technologies* (Vol. 13, No. 2, pp. 121–155).
18. Seunghyeon, L., et al., (2019). An advanced deep learning approach to real-time estimation of lane-based queue lengths at a signalized junction. In: *Transportation Research Part C: Emerging Technologies* (Vol. 109, pp. 117–136). ISSN: 0968-090X.
19. Mohammadi, M., & Al-Fuqaha, A., (2018). Enabling cognitive smart cities using big data and machine learning: Approaches and challenges. In: *IEEE Communications Magazine* (Vol. 56, No. 2, pp. 94–101). doi: 10.1109/MCOM.2018.1700298.
20. Noursalehi, P., Koutsopoulos, H. N., & Zhao, J., (2020). Machine-learning-augmented analysis of textual data: Application in transit disruption management. In: *IEEE Open Journal of Intelligent Transportation Systems* (Vol. 1, pp. 227–236). doi: 10.1109/OJITS.2020.3038395.
21. Lample, G., Ballesteros, M., Subramanian, S., Kawakami, K., & Dyer, C., (2016). *Neural Architectures for Named Entity Recognition.* arXiv preprint arXiv:1603.01360.
22. Hofmann, T., (2001). Unsupervised learning by probabilistic latent semantic analysis. *Machine Learning, 42*, 177–196. https://doi.org/10.1023/A:1007617005950.
23. Blei, D. M., & Lafferty, J. D., (2009). Topic models. *Text Mining: Theory and Applications, 10*(1), 71–93.

24. Cohen, S., (2016). Bayesian analysis in natural language processing. *Synthesis Lectures on Human Language Technologies, 9*(2), 1–274.
25. Shenhao, W., Qingyi, W., & Jinhua, Z., (2020). Multitask learning deep neural networks to combine revealed and stated preference data. *Journal of Choice Modeling, 37*, 100236. ISSN 1755–5345. https://doi.org/10.1016/j.jocm.2020.100236.
26. Yongxue Tian & Li Pan., (2015). Predicting short-term traffic flow by long short-term memory recurrent neural network. *In IEEE International Conference on Smart City/socialcom/sustaincom*, pp. 153–158, 2015.
27. Nitish Srivastava, Geoffrey Hinton, Alex Krizhevsky, Ilya Sutskever, & Ruslan Salakhutdinov, (2014). Dropout: a simple way to prevent neural networks from overfitting. *The Journal of Machine Learning Research, 15*(1), 1929–1958, 2014.
28. Giuseppe Amato, Fabio Carrara, Fabrizio Falchi, Claudio Gennaro, Carlo Meghini, & Claudio Vairo, (2016). Deep learning for decentralized parking lot occupancy detection. Expert Systems with Applications, 72.
29. Xinchen Liu, Wu Liu, Tao Mei, & Huadong Ma, (2016). A deep learning-based approach to progressive vehicle re-identification for urban surveillance. *In European Conference on Computer Vision,* pp. 869–884. Springer.
30. Jakub Sochor, Adam Herout, & Jiri Havel, (2016). Boxcars. 3d boxes as CNN input for improved fine-grained vehicle recognition. *In Proceedings of the IEEE Conference on Computer Vision and Pattern Recognition*, pp. 3006–3015.
31. Arief Koesdwiady, Ridha Soua, & Fakhreddine Karray, (2016). Improving traffic flow prediction with weather information in connected cars: a deep learning approach. *IEEE Transactions on Vehicular Technology, 65*(12), 9508–9517.
32. Wenhao Huang, Guojie Song, Haikun Hong, & Kunqing Xie, (2014). Deep architecture for traffic flow prediction: deep belief networks with multitask learning. *IEEE Transactions on Intelligent Transportation Systems*, *15*(5), 2191–2201.
33. Hao-Fan Yang, Tharam, S., Dillon, & Yi-Ping Phoebe Chen, (2017). Optimized structure of the traffic flow forecasting model with a deep learning approach. *IEEE transactions on neural networks and learning systems*, *28*(10), 2371–2381.
34. Yanjie Duan, Yisheng Lv., Yu-Liang Liu, & Fei-Yue Wang, (2016). An efficient realization of deep learning for traffic data imputation. *Transportation Research Part C: Emerging Technologies, 72,* 168–181.
35. Dong Wang, Junbo Zhang, Wei Cao, Jian Li, & Yu Zheng, (2018). When Will You Arrive? Estimating Travel Time Based on Deep Neural Networks. *In Proceedings of the Thirty-Second AAAI Conference on Artificial Intelligence.*
36. Statista.com, (2020). Online: https://www.statista.com/statistics/330695/number-of-smartphone-users-worldwide/.
37. Ouyang, X., Zhang, C., Zhou, P., Jiang, H., & Gong, S., (2016). DeepSpace: An Online Deep Learning Framework for Mobile Big Data to Understand Human Mobility Patterns. *arXiv e-prints.*
38. Zhao, L., Li, Q., Zhang, Y., Wang, H., & Du, X. (2019). Integrating the Continuous Wavelet Transform and a Convolutional Neural Network to Identify Vineyard Using Time Series Satellite Images. *Remote Sensing, 11*(22), 2641.
39. Dogan, E., & Akgüngör, A. P., (2013). Forecasting highway casualties under the effect of railway development policy in Turkey using artificial neural networks. *Neural Comput. Appl., 22*, 869–877.

40. Budalakoti, S., Srivastava, A. N., & Otey, M. E., (2009). Anomaly detection and diagnosis algorithms for discrete symbolsequences with applications to airline safety. *IEEE Trans. Syst. Man Cybern. Part C Appl. Rev.*, *39*, 101–113.
41. Akgüngör, A. P., & Dogan, E., (2009). An artificial intelligent approach to traffic accident estimation: Model development and application. *Transport, 24*, 135–142.
42. Huang, W., Song, G., Hong, H., & Xie, K., (2014). Deep architecture for traffic flow prediction: Deep belief networks with multitask learning. *IEEE Trans. Intell. Transp. Syst.* 2014, 15, 2191–2201.
43. Jiang, H., Zou, Y., Zhang, S., Tang, J., & Wang, Y., (2016). Short-Term Speed Prediction Using Remote Microwave Sensor Data: Machine Learning versus Statistical Model. *Math. Probl. Eng., 2016,* 9236156.
44. Ledoux, C., (1997). An urban traffic flow model integrating neural networks. *Transp. Res. Part C Emerg. Technol.*, *5*, 287–300.
45. Lv, Y., Duan, Y., Kang, W., Li, Z., Wang, F.Y., (2014). Traffic Flow Prediction with Big Data: A Deep Learning Approach. *IEEE Trans. Intell. Transp. Syst.*, *16*, 865–873.
46. More, R., Mugal, A., Rajgure, S., Adhao, R. B., & Pachghare, V. K., (2016). Road traffic prediction and congestion control using Artificial Neural Networks. *In Proceedings of the 2016 International Conference on Computing, Analytics and Security Trends (CAST),* Pune, India, 19–21 December 2016; pp. 52–57.
47. Wu, Y., Tan, H., Qin, L., Ran, B., & Jiang, Z., (2018). A hybrid deep learning based traffic flow prediction method and its understanding. *Transp. Res. Part C Emerg. Technol., 90*, 166–180.
48. Theofilatos, A., Yannis, G., Kopelias, P., & Papadimitriou, F., (2016). Predicting Road Accidents: A Rare-Events Modeling Approach. *Transp. Res. Procedia, 14*, 3399–3405.
49. Król, A., (2016). The Application of the Artificial Intelligence Methods for Planning of the Development of the Transportation Network. *Transp. Res. Procedia, 14*, 4532–4541.
50. Xu, T., Wei, H., Wang, Z. D., (2009). Study on continuous network design problem using simulated annealing and genetic algorithm. *Expert Syst. Appl., 36* Pt 2, 2735–2741.
51. Ceylan, H., & Bell, M. G. H., (2004). Traffic signal timing optimization based on genetic algorithm approach, including drivers' routing. *Transp. Res. Part B Methodol., 38,* 329–342.
52. Ulusoy, Ü., Sivrikaya¸ Serifoglu, G., & Bilge, F., (1997). A genetic algorithm approach to the simultaneous scheduling of machines and automated guided vehicles. *Comput. Oper. Res.*, *24*, 335–351.
53. Karoonsoontawong, A., & Waller, S., (2006). Dynamic Continuous Network Design Problem: Linear Bilevel Programming and Metaheuristic Approaches. *Transp. Res. Rec. J. Transp. Res. Board, 1964*, 104–117.
54. Xu, T., Wei, H., & Hu, G., (2009). Study on continuous network design problem using simulated annealing and genetic algorithm. *Expert Syst. Appl.*, *36* Pt 1, 1322–1328.
55. Kirkpatrick, S., Gelatt, C. D., & Vecchi, M. P., (1983). Optimization by Simulated Annealing. *Science, 220*, 671–680.
56. Yaseen, S. G., & L-Slamy, N. A., (2008). Ant Colony Optimization. *Int. J. Comput. Sci. Netw. Secur.*, *8*, 351–357.

57. Dorigo, M., Gambardella, L. M., Birattari, M., Martinoli, A., Poli, R., & Stützle, T., (2008). LNCS 4150—Ant ColonyOptimization and Swarm Intelligence. *In Proceedings of the 6th International Conference (ANTS 2008)*, Brussels, Belgium, 22–24 September 2008.
58. Timmis, J., Neal, M., & Hunt, J., (2000). An artificial immune system for data analysis. *Biosystems, 55*, 143–150.
59. Lucic, D.B., & Teodorovic, P., (2001). Bee system: Modeling combinatorial optimization transportation engineering problems by swarm intelligence. *In Proceedings of the TRISTAN IV Triennial Symposium on Transportation Analysis*, Azores, Portugal, 13–19 June 2001.
60. Lucic, D., & Teodorovic, P., (2002). Transportation modeling: An artificial life approach. *In Proceedings of the 14th IEEE International Conference on Tools with Artificial Intelligence (ICTAI 2002)*, Washington, DC, USA, 4–6 November 2002.
61. Lucic, D., & Teodorovic, P., (2003). Computing with bees: Attacking complex transportation engineering problems. *Int.J. Artif. Intell. Tools. 12*, 375–394.
62. Lucic, D., & Teodorovic, P., (2003). Vehicle routing problem with uncertain demand at nodes: The bee system and fuzzy logic approach. In Fuzzy Sets Based Heuristics for Optimization; Springer: Berlin/Heidelberg, Germany, pp. 67–82.
63. Qureshi, M. F., Shah, S. M. A., & Al-Matroushi, G. I. G., (2013). A Comparative Analysis of Multi-Criteria Road Network. Eur. Cent. Res. Train. Dev. UK, 27–47. Available online: http://www.eajournals.org/wp-content/uploads/A-Comparative-Analysis-of-Multi-criteria-Road-Network.pdf.
64. Harcourt, P., Route Optimization Techniques: An Overview. Available online: https://www.ijser.org/researchpaper/route-optimization-techniques-an-overview.pdf.
65. Bagloee, M., Sarvi, S. A., & Patriksson, M., (2017). A hybrid branch-and-bound and benders decomposition algorithm for the network design problem. *Comput. Civ. Infrastruct. Eng., 32*, 319–343.
66. Li, X., Shi, X., He, J., & Liu, X., (2011). Coupling simulation and optimization to solve planning problems in a fast-developing area. *Ann. Assoc. Am. Geogr., 101*, 1032–1048.

CHAPTER 4

Smart Mobility Using AI and Intelligent Agents

SHRDDHA SAGAR,[1] NILANJANA PRADHAN,[2] and SUPRIYA KHAITAN[3]

[1]*Professor, School of Computing Science and Engineering, Galgotias University, Greater Noida, Uttar Pradesh, India*

[2]*Research Scholar, School of Computing Science and Engineering, Galgotias University, Delhi-NCR, India.*

[3]*Assistant Professor, Pillai College of Engineering, University of Mumbai, Mumbai, Maharashtra, India*

ABSTRACT

The way we live our lives and carry out our daily tasks has altered dramatically in the first few months of 2020. Although widespread deployment of futuristic robot axis and self-driving commercial vehicles is still a long way off, it has pushed the use of artificial intelligence (AI) in a variety of industries. Collecting, storing, analyzing, exploiting, and disseminating multi-source data has become easier and less expensive thanks to emerging technologies for a connected vehicle-infrastructure-pedestrian ecosystem, as well as big data. The idea of an agent has been adopted by both mainstream computer science and AI. In order to increase overall system performance, the linked environment brings new techniques to flexible control and administration of transportation systems in real time. Given the benefits of a connected world, we must understand how the current intelligent transportation system (ITS) may be adapted to it. ITSs sensing,

Smart Mobility and Intelligent Transportation Systems for Commercial and Hazardous Vehicles.
T. Poongodi, Naga Swetha Pasupuleti, and Balamurugan Balusamy (Eds.)

communication, data dissemination, and traffic control are only some of the technologies that have been integrated. Data gathering, data analysis, and data/information acquisition are all steps in the data collection process transfer are three key components of any ITS in order for it to perform its function(s). The chapter's major goal is to present the most significant conceptual and practical issues related to the creation of intelligent agents, as well as smart mobility employing AI.

4.1 IMPORTANCE OF AI OVER MOBILITY

Artificial intelligence (AI) is a buzzword these days. Many items and services labeled as such in the automotive sector today, however, are actually depending on advanced analytics (derived from traditional algorithms) to deliver certain features—for example, in production, and predictive maintenance. AI approaches are not new but are already existing since the 1950s. However, as machine learning (ML) and deep learning (DL) become more prevalent, AI is a relatively new technology with a wider domain of applications. The cloud's ability to store a lot of data, more powerful computers, and improved algorithms and training techniques have all helped this process.

Despite these advancements, the AI disruption in the automotive industry is still in its early stages. AI applications that are currently available are very limited—they can outperform humans in relatively specialized activities. Furthermore, the scope and form of AI technology differ greatly; Narrow AI, for example, includes both standard navigation systems and self-driving vehicles that handle one gigabyte of data per second, approximately a million times the amount of data now available to navigation systems [1].

Despite all of the buzz about what AI might be capable of, is it all hype, or is it an essential technology that companies must master? AI advancements on a higher level could provide a significant competitive advantage, which is especially true in the automotive and mobility industries. Machine-learning and deep-learning approaches are used by AI systems in the automobile industry to improve or build new applications. By the implementation of ML algorithms makes AI more acceptable in automotive and mobility [2]. The main domain in which ML can provide the solutions are more complex situations, deal with a vast number of potential cases that explicit programming can't handle and that can't be handled without explicit instructions, enhance over time by learning from unstructured data from previously unforeseen scenarios.

In mobility, ML will not be an optional component of AI; For decades to come, it will be the technological foundation and a significant source of competitive advantage. The automotive and transportation industries will be significantly impacted by ML, more than simply increasing productivity, it will enable the creation of entirely new goods and value pools [3]. In a mobility context, machine-learning technology is so intricate that it will need the establishment of new structures within and surrounding the mobility ecosystem. In the new industry landscape, there are numerous stakeholders, as well as the technology that supports it, will demand totally new commitments, processes, and working ways.

There will be a range of ML-enabled systems in and around the automobile, some embedded, some external, includes periodic communication and updates to a back end, as well as those that are cloud-based and pushed to the car in real time [4]. Local circumstances and legislation have an impact on all of these systems. They also rely on a variety of stakeholders (including third parties) as well as different technology restrictions.

4.1.1 BENEFITS OF AI

Artificial intelligence (AI) is pushing the limits of machine capabilities. This cutting-edge technology allows machines to operate autonomously, allowing them to complete iterative tasks more efficiently [5]. AI helps to create the next-generation workplace, which thrives on seamless communication between enterprise systems and employees. As a result, human resources are not rendered obsolete; rather, rising technology augments their efforts. In reality, AI allows businesses to free up resources to focus on higher-level projects [6]. The major benefits of AI are as follows:

- AI shortens the amount of time needed to finish a task. It makes it possible to multitask and reduces the load on the available resources.
- AI accepts the execution of previously complicated activities without incurring significant costs.
- AI is available without interruption or downtime 24 hours a day, 7 days a week.

- AI enhances the talents of people with a variety of skills.
- A wide range of sectors might adopt AI, which has a huge market potential.
- AI facilitates decision-making by enhancing its speed and intelligence.

4.1.2 CHALLENGES OF AI

Though there are many benefits of implementing AI for smart mobility but still it faces some challenges which are as follows [7]:

1. **Methodologies based on Data:** Developing AI tools, like as machine learning, to process and interpret heterogeneous data from various sources in order to improve operations and affect traveler behavior, where the interpretability of outcomes must be considered.
2. **Algorithms for Intelligent and Hybrid Planning:** To handle planning and decision challenges in the mobility area, intelligent algorithms, including optimization approaches and learning, are being developed.
3. **AI with a Human Focus:** Intelligent interactive information presentations, AI-induced persuasive technology, and personalized interaction to maximize user satisfaction are among the AI techniques being developed to understand and predict human choice behavior and persuade people to make better choices (efficient, environmentally friendly, etc.).
4. **Simulation of Intelligent Systems:** Develop AI approaches like agent-based simulation to analyze the design and performance of mobility concepts, including decision models and the roles of various players.

4.2 AI IMPACTING MOBILITY

There are several mobility alternatives for AI, which has the potential to be a crucial source of competitive advantage. Automobile OEMs, on the other hand, are well positioned to benefit from AI advances due to the embedded control point. Furthermore, the majority of customers expect automakers to be at the forefront of autonomous technology development. To bring them to market, rather than IT companies—interestingly [8]. Rather than tech businesses, car manufacturers are likely to spearhead the development of autonomous functions:

1. **Smart Grid Management:** The batteries in electric cars must be charged on a regular basis. Many consumers, however, are unaware that the energy stored in electric car batteries may be sold back to the grid. AI can help you figure out when it's best to charge your electric vehicle and when it's best to use it for V2G "un-charging." Drivers can save money while simultaneously enhancing the grid's efficiency and stability with smart grid management.

2. **Transportation System:** Users may arrange excursions utilizing various modes of transportation rapidly with mobility-as-a-service (MaaS). Using mobile phones or other linked devices as personal devices, commuters can utilize MaaS to organize, manage, and pay for transportation. When AI is implemented in MaaS systems, significant benefits emerge, such as fully autonomous driving and smart tracking.

3. **Monitoring of the Driver:** A driving monitoring system employs cameras to check the driver's attentiveness. The device recognizes the driver and determines his or her degree of concentration. The system alerts a driver if they are inattentive. AI-based technologies allow drivers to customize vehicle settings such as seat position, temperature, and mirrors to their preferences.

4. **Self-Driving Vehicles:** Autonomous vehicles (AVs) are slowly but surely finding their way into the transportation sector, despite widespread mistrust. While the majority of self-driving car businesses are currently working on pilot programs to ensure passenger safety, others are moving forward. Vehicles have already been deployed on public roads by a number of companies. The brain of a self-driving automobile is made up of computer vision and DL technology. These systems process all of the data from the sensors and make context-dependent decisions.

5. **Manufacturing:** As a result of AI, the way cars are made is changing as well. The H-VEX (Hyundai Vest Exoskeleton) wearable robot, for example, was recently unveiled by Kia. The robot is used to optimize vehicle assembly lines as well as to protect workers in the industry. Another example is automated guided vehicles (AGVs), which are designed to transport materials without the need for human intervention. AI-enabled welding and painting robots can detect production faults and change their methods accordingly.
6. **Smart Cities:** AI-based technologies have a significant impact on the expansion and evolution of cities. AVs, for example, may help to initiate a de-urbanization trend. Commutes are cheaper, faster, and safer with AVs. People can live outside of the city and use self-driving cars to get to work. Because they aren't driving, they can be completely productive during the travel. They can work while the AI drives the vehicle during their commute.

4.3 SMART CITY GRID MANAGEMENT

Connecting interfacing goods across available networks has become increasingly popular with the advancement of smart gadgets. Smart gadgets are designed to share/access information with/from other devices so that intelligent decisions may be made using that information. Integrated smart devices are essential for this to happen, and they must be as seamless as possible [9]. The United Nations estimates that 70% of the world's population will reside in cities by 2050 as a result of people migrating there to improve their living conditions. A number of challenges are unavoidably brought on by urbanization, including resource shortages, inefficient energy and waste disposal, deteriorating infrastructure, and poor human health. A decade ago, experts proposed the concept of smart cities as an ideal solution to resolving urbanization challenges [10]. Information and communication technology (ICT) used to improve the quality of life in cities and among its residents is referred to as "smart technology." Smart cities aren't just places where information and communications technology (ICT) is integrated into urban surroundings in fact 'Smart city' has many definitions, but the most widely accepted one is that it's connecting the environment of physical, social, business, and information technology

(IT) infrastructure for elevating the intelligence of cities by balancing the demand and supply of various functionalities; Cities consume 75–80% of global energy, resulting in 80% of global greenhouse gas emissions [11]. Apart from fossil fuel constraint and urban population growth, renewable energy sources can best meet the recent global energy demands of sustainable smart cities because of their renewable nature. In order to create a city of the future that is both environmental friendly and economically viable, we need smart grids, which are modernized versions of traditional energy infrastructure. Smart gadgets are increasing the demand for interfacing items to be connected through available networks [12]. Smart gadgets are designed to share/access information with/from other devices so that intelligent decisions can be made using the information they provide. Integrated smart devices are essential for this to happen, and they must be as seamless as possible. Contrary to popular belief, 70% of the world's population, according to UN predictions, will reside in cities by 2050. As we know, urbanization brings with it a host of new problems, including resource scarcity, pollution, deterioration of public infrastructure, and poor health. A decade ago, experts proposed the concept of smart cities as an ideal solution to resolving urbanization challenges. Using ICT to improve the quality of life in cities and among its residents is described as being smart in this section as the desire to do so from several angles (ICT). Smart cities aren't just places where information and communications technology (ICT) is integrated into urban surroundings, it can mean many things, but one frequent definition is "the integrated environment of the physical, social, business, and IT infrastructure for upgrading the city's intelligence" [13]. Cities use 75–80% of the world's energy, resulting in 80% of the emissions of all greenhouse gases. Traditional energy infrastructures are being modernized to create smart grids, which are fundamental to the idea of a sustainable future city.

4.3.1 INNOVATIVE MODES OF SMART TRANSPORTATION

Cities are ever-changing entities. Transportation is essential for the survival of cities. Land and transportation planners, modelers, mathematicians, and physicists are increasingly studying transportation's complex impact on cities' dynamic health and resilience. Traffic congestion is estimated to cost Americans 4.8 billion hours each year, according to data from the transportation sector. In 2013, there were 5.7 million car accidents reported to the authorities. Around 3.9 billion gallons of petroleum were thrown away last year alone. By making transportation networks smarter, most of this loss of environmental and human values can be averted. About 80% of automobile collisions will be avoided with the use of one of the car technologies discussed in this chapter [14].

Large-scale spatiotemporal data gathering on human mobility is now technically and economically feasible thanks to advances in information and communication technologies (ICTs). Mobile phone and smart card location data, taxi schedules, and social media posts are a few examples of spatiotemporal big data. Large-scale spatiotemporal big data sets like this enable us to see how many people move around on a daily basis in incredible detail. It is possible to gain a better understanding of people's activity-travel behaviors and interactions with the transportation environment and policy using such large data sets. This creates opportunities to create new, more advanced models and algorithms that will help to improve transportation systems in terms of intelligence and safety. As China's economy continues to expand, there has been a significant increase in the number of cars on Chinese city streets [15]. As a result of this change, urban transportation has been severely strained. To deal with this problem, traffic management has steadily integrated urban intelligent transportation systems (UITS) with intelligent characteristics. Big data is used in the city's intelligent transportation system (ITS). It can handle data more effectively and perform better. In addition to road and other video surveillance data, the enormous data set contains information from many forms of traffic monitoring. Detecting traffic movement on city streets and highways, meteorological data, urban public transportation, and vehicle satellite location data are just a few examples. There are many different forms of traffic data, and the collection is quite large. Following in-depth studies of the market, it has been discovered that while there are a number of systemic products for traffic management available both domestically and internationally, there are still some drawbacks. These include single-system functions such as missing integration and backward technology. Large-scale spatiotemporal data collection on human mobility is now technically and financially viable thanks to the development of information and communication technologies (ICTs). Space-time big data includes anything from mobile phone location to credit card transactions to taxi routes and social media posts. We can use huge data like this to better understand people's activity-travel behaviors and connections with the environment and policy of transportation [16]. This creates the potential for the creation of cutting-edge models and algorithms that will support the intelligent and secure functioning of transportation networks. The urban ITS can function better and handle data more efficiently with the use of big data technologies. The large data set includes all sorts of traffic monitoring at the same time, such as traffic flow monitoring on city streets and highways, as well as road and other video surveillance data. Examples include information about the weather, city transit, and car locations as determined by satellite. There are some traffic management systems that exist, but market research and analysis. Unintegrated functions of separate systems and other issues remain as obstacles, and backward technology, which are usually observed in constructing application systems that distribute mass data, as well as a lack of effective traffic data integration and low utilization [17].

4.3.2 DRIVING RECORD SURVEILLANCE

An overtired driver can occur at any time when operating a moving vehicle (such as a car, bus, or truck). This may help to avoid an accident involving that vehicle by alerting the driver before the two vehicles collide. It's only the sleepy driving problem that gets more attention; the other two are less of a concern. Brain wave sensors and eye monitoring are used in the current sleepy driver technologies to detect drivers who are falling asleep behind the wheel. Different methodologies can be utilized to detect the eye activity using the SURF Cascade approach (based on the VJ algorithm) and brain wave analysis with a sensor. Installed in the seatbelt is a heart rate monitor to monitor heart activity. Perspiration on the palms monitors alcohol intake using sensors that detect alcohol scent.

To improve accuracy, additional alcohol sensors may be placed near the driver. When a driver is ill or sleepy, the automobile automatically slows down and comes to a complete stop with the flashing light/alert lights on, using data from the previously described sensor modules.

Due to an increase in the incidence of vehicle accidents, active safety systems have been studied and designed in current cars. Several sensors for measuring acceleration, wheel velocities, and vehicle yaw have been installed for this goal. Although a variety of important factors, including slip angle and frictional forces, can be detected by sensors, it is not practical to do so with any degree of accuracy. Due to the importance of these measurements in creating a trustworthy control system, active safety system design strongly depends on the estimation of friction coefficient and frictional forces. The linearized estimating technique may also produce huge approximation errors because the system is highly nonlinear. The friction coefficient and lateral and longitudinal frictional forces are estimated using a nonlinear tire model and an unscented Kalman Filter-based estimation approach.

Over the course of 2014, 846 persons lost their lives as a result of drowsy driving (National Center for Statistics and Analysis (2017); Speeding data in 2015). According to recent estimates, driver inattention is responsible for 6% of all collisions and 21% of fatal collisions. Crash reports are notoriously difficult to get [22, 23]. Driving when fatigued is something that 28% of drivers have admitted to doing in the last month, according to a new survey [24]. Collisions caused by drowsy driving may be reduced with the aid of in-car driver state systems. These systems include a detecting system as well as a countermeasure. When a driver is excessively tired, signs such as irregular steering and lane departures can be observed.

The "coffee cup" interface is the most popular, in which an instrument panel or infotainment system displays a coffee cup icon and an audible alarm sound. Existing countermeasure systems offer messages and warnings as forms of feedback. As an alternative to sending a binary signal, some systems employ a "attentiveness" scale to describe the current status. However, there are other systems that use vibrations as well as images displayed on the dashboard or steering wheel [18].

4.4 SELF-DRIVING VEHICLES

Studies on tired driver feedback countermeasures are scarce. Researchers found that simple sound warnings like these improved lane keeping. Visual warnings improved lane retention over time compared to no feedback. In 4-hour trips, it was tested the effectiveness of combined auditory/visual alerts for drowsy drivers using dashboard-mounted LED lights and an audio tone. The vibration of the seat belt caused by drowsiness increased lane maintenance and reduced driver weariness. Combining different sorts of warning modalities has been shown to improve driving performance in previous research. While most commercially available warning systems employ one of these two approaches, there are a few that use both for example, when a driver shows signs of tiredness, an espresso cup icon appears, a scale that indicates the level of driver alertness. Several modalities and interaction types were included in the current study's countermeasures in order to accurately portray current in-vehicle countermeasures (discrete, staged).

Numerous colleges, research organizations, automobile manufacturers, and companies from other industries have been investigating and developing self-driving automobiles since the mid-1980s AVs (also known as driverless vehicles or driverless automobiles). Many self-driving automobile research platforms have been established during the past two decades, including the Navlab mobile platform, which was developed by the University of California at Berkeley, the ARGO car developed by the Universities of Pavia and Parma as well as the VaMoRs and VaMP automobiles developed by UBM. As a way to encourage self-driving car development, DARPA has organized three tournaments in the recent decade [19]. Two noteworthy contests were place in 2004: the DARPA Grand Challenge, which required self-driving cars to travel 142 miles in 10 hours through the Mojave Desert. Within the first several miles, all of the competing vehicles had been destroyed. A 132-mile route over lowlands, dried lake beds, and mountain ranges was part of the 2005 DARPA Grand Challenge. More than 100 quick left- and right-turns were included, along with three small tunnels. Four cars triumphed out the other 22 competitors by completing the course in the allocated time. In the end, Stanford University's Stanley 6 automobile took first place, with Sandstorm and Highlander from Carnegie Mellon University coming in second and third, respectively. As part of the DARPA Urban Challenge, self-driving automobiles were to complete a 60-mile route in six hours using human drivers and AVs in June 2007 at the formerly George Air Force Base in California [25]. The vehicles have to abide by state traffic regulations in California. Out of the 11 cars that started, six finished the track in the given time. Boss, a Carnegie Mellon University car, took top place, followed by Junior, a Stanford University car, and Odin, a Virginia Tech car, who finished third overall. AVs have been lauded as technological advances despite the fact that the roadblocks they faced were significantly smaller than those encountered during normal traffic. Numerous self-driving car competitions and testing have been conducted since the DARPA challenges. Relevant examples include ELROB, the Land Trial Robot for European, Challenges for Future Smart Vehicle, Competition for Automated, and Visual Lab Intercontinental Automated Challenge. They all took place between 2006 and the present year, and they all took place in Europe. Simultaneously, academic and industrial research into self-driving cars has advanced around the world. Stanford University is a significant academic institution engaged in research on AVs. Other notable universities conducting this study include Mellon University of Carnegie, Massachusetts Institute of Technology (MIT), Technology of Virginia, the FZI Research Center for Information Technology, and the University of Ulm. Google, Uber, Baidu, and Lyft are well-known companies, as are Tesla, Nvidia, Aurora, Zenuity, Daimler, and Bosch, as well as Argo AI [20]. Other noteworthy companies include Toyota, Ford, and Volvo, as well as Renesas, Almotive, and AutoX. Even while China, Brazil. Most of the research on autodriven automobiles research has been done in the US, European Countries, and Asian Countries, while other countries have made significant contributions. Among other notable instances, self-driving automobile platforms in Brazil include the CADU at UFMG, the CA-RINA at the University of So Paulo, and the IARA at the University of Espirito Santo (UFES). IARA, the country's first autonomous vehicle, traveled tenths of a kilometer on highways and city streets by itself. The Society of Automotive Engineers International (SAE) claims that, self-driving cars can have an autonomy level ranging from 0 to 100% (in which the autonomous system sends warnings and may intervene for a brief period of time but has no long-term causation) to five (in situations where the autonomous system issues warnings and may intervene for a limited time, but has no long-term impact) depending on the level of human involvement and attention required.

4.5 TRAFFIC MANAGEMENT

The problem of traffic congestion is caused by an increase in the number of automobiles on the roadways. In reality, there is a bottleneck in the time distribution for passing cars at intersections, and vehicles tend to be queued up along the road in one way but not so much in the other. In this situation, by creating an intelligent traffic signal system, we can reduce traffic congestion and increase the efficiency of managing traffic junctions.

Over the last two decades years of development, refinement, and practical use, current systems have shown a number of flaws, including [16, 17]:

- Shortcomings in the accuracy and scope of using traffic prediction models.
- Inadequate ability to handle a wide range of traffic circumstances, especially those that are very congested.
- Inadequate capacity to deal with traffic situations separated in space or time from a local control choice that is still influencing or being influenced by it.
- Constrained ability to adapt to changing environmental conditions (beyond apparent differences in need for transportation).
- Instead of isolated intersections, adaptive systems are now linked to network traffic control or corridor traffic control. Because of this, the solution for a single intersection is either inadequate or unusable [19].
- Gartner has emphasized on the need of developing new concepts based on intelligent systems [17, 18].

The outdated traffic management system clearly has to be greatly improved by incorporating intelligence to enable dynamism and adaptivity in light of its flaws. As a result, there will be less congestion on the roads, which will cut down on travel time and pollution. A number of decades ago, the traditional traffic management system was developed. At the time, the number of cars was quite modest, and the traditional system was adequate to efficiently control traffic with the existing technology. Designing, optimizing, implementing, and modifying the downtown traffic management systems necessitates a significant amount of work and knowledge.

Changing surroundings may not necessarily result in changes in traffic scenario for a variety of reasons. Accustoming a traffic control device is an expensive and time-consuming process. For short-term traffic patterns, a traffic control device not modified or updated. The capacity of traffic management systems to respond to changes in traffic patterns is critical to their success. The more this capacity to react gets ingrained in the traffic management unit, the better it will be able to respond to changes in traffic circumstances.

The responsive control system should be familiar enough with the environment to manage any kind of unexpected changes in traffic flow, accident is one such example. An Intelligent Traffic management system must be able to handle these unexpected changes. To minimize the delay caused by unexpected delay and increase throughput, changing traffic lights and coordinating operation between each signal can be optimized by using intelligent systems. We need a predictive and pro-active, real-time traffic control system for urban traffic control; predicting what will happen in the following few minutes [16].

The control plans employed in such a system are based on triggered traffic circumstances and are updated regularly, potentially even during a cycle, and are optimized by some type of performance criterion. Responsive plan generation is a frequently debated but seldom realized notion in urban traffic control. The main concept is that existing signal-plan generating technologies (e.g., UTOPIA-SPOT or SCOOT) produce reasonable signal-plan judgments under various situations.

Abbas & Tufvesson [26] introduced two sorts of agents, one central and the rest local. The local agent at each junction is in charge of updating the light signals. The proposed local agent gets the vehicle queue length from sensors placed on each side of the junction. Then, execute a series of control instructions to update the timing of the green light on either side of the junction. While the central agent is perceiving the state of the local agents in various places, it analyzes the received data and sorts the nodes in various locations before sharing the announcement at each node in the CITY road map that describes the CITY road map at the time.

ITS are a relatively new technology that intends to provide novel services connected to various traffic management and transportation systems. Furthermore, it helps users to be better informed and make more coordinated, intelligent, and safe use of transportation networks. ITS attempts to improve transportation efficiency by minimizing traffic congestion. This technology provides the user with previous knowledge regarding seat availability, real-time running information, traffic, and so on, reducing travel time and improving the user's safety and comfort.

For effective utilization of traffic infrastructure, congestion control, safety measures and most important congestion control, the usage of Intelligent Traffic System is widely acknowledged and employed in many nations. Intelligent Traffic Systems has expanded into a multidisciplinary conjunctive field of endeavor as a result of its enormous potential, and many organizations around the world have created solutions to meet the demand for ITS applications.

One such city is Glasgow, where the Intelligent Transport System regularly informs commuters about public buses, timetables, and the amount of time needed to travel to a particular location. Seat availability, the bus's current location, the bus's next location, and the bus's passenger density are all factors to consider. The sensors are installed in all city's bus, in case of any delay. The Intelligent Traffic System's entire application is based on requirement analysis, data collecting, and the application of analytic results to traffic management operations, controls, and research projects where location is a key factor. In the application of Public Transportation, Rural Transportation, Management of Traffic, System for Smart Vehicle Control, Advanced Traveler Information systems, roadside messages, GPS updates, automatic traffic prioritization signals, sensors, information processors, communication systems, and advanced commercial vehicle operations systems all play a crucial role.

The Traffic Management Center is an essential component of Intelligent Traffic System. All data is collected and evaluated for handling real time traffic situations by transportation authorities. The Traffic Management Center's efficient and well-managed operations depend on the automated data collection of exact location data, analysis of that data to produce accurate information, and transmission of that information to users. Let's take a closer look at the entire procedure as shown in Figure 4.1.

1. Data Gathering: It is important to gather data in a timely manner for real-time monitoring. As a consequence, data is collected here via a range of hardware devices, which serve as the foundation for future Intelligent Transport System operations. These devices include GPS-based automatic vehicle locators (AVLs), automatic vehicle identifiers (AVIs), sensors, digital cameras, and so on. The gear simply captures information like traffic counts, surveillance, journey speed and length, position, vehicle weight, and delays, among other things. Computers, which are frequently found at data collecting centers and store vast amounts of data for later analysis, are connected to these physical devices.
2. Transmission of Data: It is of utmost importance to rapidly transmit the information for deployment competency in Intelligent Transport System. Data transmission component of Intelligent Transport System consists of transmitting gathered data from the

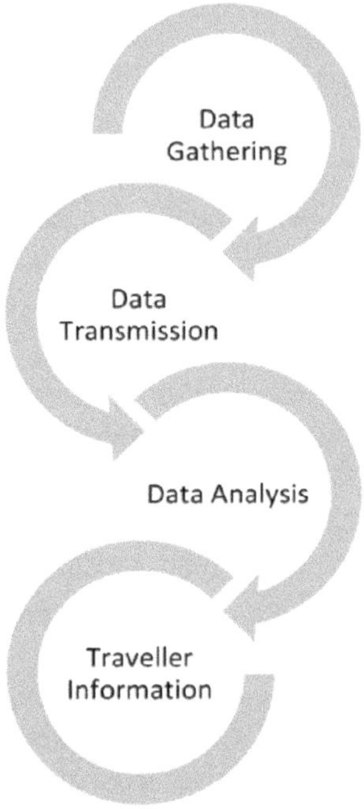

FIGURE 4.1 Traffic management system.

field to Traffic Management control server and then analyze the information and send it back to travelers. Travelers receive traffic related notifications via intelligent devices like smartphone apps, SMS, etc. Two more methods of communication are continuous air interface long and medium range (CAILM), which uses cellular connectivity, and dedicated short-range communication (DSRC), which uses radio.

3. Analyze the Data: The information gathered and received by Traffic Management Control is handled in a different method. Data is cleaned, synthesized, errors are corrected, and inconsistencies are removed. For analysis, the information is updated and consolidated. This enhanced aggregate data is then used to estimate traffic situations, which are then made available to customers in order to provide them with relevant information.
4. Information for Travelers: This is meant to notify users about current information, such as travel time, delays, traffic incidents,

route changes, journey time, detours, and work zone conditions. Internet, SMS, automatic message signs, radio for motorway advisories, and mobile phones are among the electronic technologies used to deliver this information.

4.6 INTELLIGENT MANUFACTURING

Intelligence manufacturing is the combination of the intelligence of human, machines, and process, to impact the economics of manufacturing. Intelligence manufacturing is used to optimize the resources used for manufacturing. It is also used to enhance the business value and safety. The manufacturing process creates lots of waste, with the help of intelligent manufacturing process the waste can be reduced on both back-office operations and floor. To achieve this the manufactures, use latest devices, data analysis tools, machine-to-machine communication for its production lines and facilities [21].

Most manufacturers in construction equipment, agriculture, heavy trucks, automotive, etc., are moving towards intelligent manufacturing solutions. This helps the manufactures to stay ahead in competition. Intelligent operation, control, planning, and maintenance are the primary elements of an intelligent manufacturing system as shown in Figure 4.2.

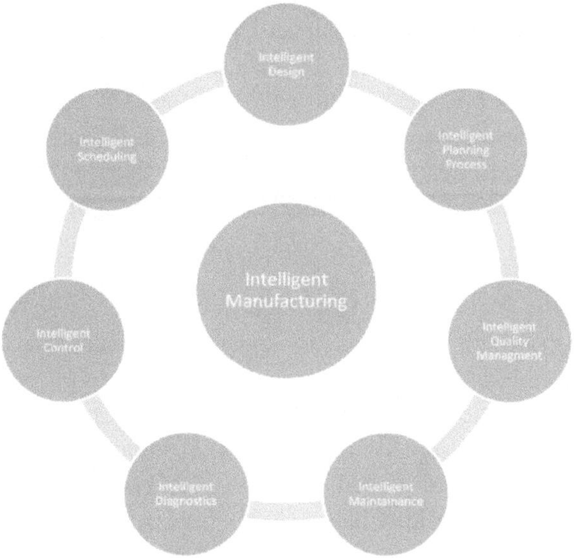

FIGURE 4.2 Intelligent manufacturing.

Monitoring the actions inside the production system and its environment increases the system reliability and prevents defects and failures. Intelligent Manufacturing depends on three paradigms Figure 4.3.

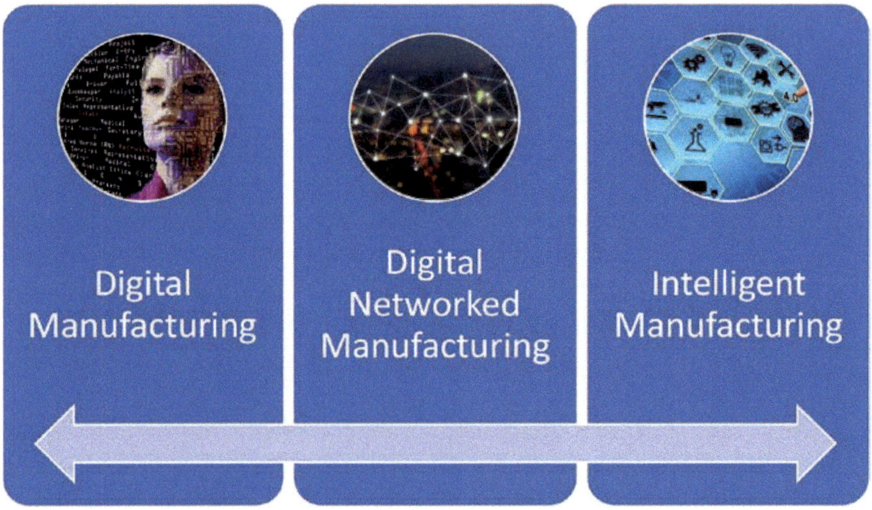

FIGURE 4.3 Paradigms of intelligent manufacturing.

1. **Manufacturing in the Digital Age:** It is the first generation of intelligent manufacturing and its foundation, Digital simulations of the factory architecture, product design, machinery utilization, and labor are created in order to create the optimum value chain that minimizes costs, improves product quality, and makes the best use of resources. Because everything is done digitally, the time necessary for making items tailored to consumer needs is reduced.

2. **Digital-Networked Manufacturing:** The usage of the internet in the computerized production system is integrated in the second generation of intelligent manufacturing. The internet network integrates ideas, processes, and data across numerous industrial units, allowing for collaborative manufacturing. Networking technology enables collaborative R&D across organizations, links firms in the same industry (horizontal integration), connects enterprises operating at various stages of the manufacturing process

within the same industry (vertical integration), and connects users to enterprises. The first stage in transforming the production process into a user-centric one is user engagement.

3. **New-Generation Intelligent Manufacturing:** The combination of AI with digital and networking technologies has resulted in key advancements in IMS. The incorporation of AI into industrial processes has the potential to radically change production systems and eliminate the requirement for human intellect. AI empowers next-generation intelligent manufacturing to conduct R&D and develop new processes, designs, products, and business models without the need for human participation. Because robots are quicker than people, AI will not only cut the time required for manufacturing but will also minimize the time required for innovation and ideation.

4.7 CONCLUSION

An examination of how AI can be used for different types of problems related to transportation. As our cities and transportation systems grow more instrumented, the spectrum of applications is anticipated to increase, giving crucial data for the creation of AI applications. The study focuses on several application areas, including autonomous cars, public transportation, disruptive urban mobility, automated incident detection, traffic management, and control, which are anticipated to have a greater impact in future cities and forecasting future traffic status. It demonstrates how AI can be applied to problems including rising transport demand, CO_2 emissions, safety concerns, and energy waste. Understanding what an agent is and how the idea of an agent may be formalized will help, how suitable agent architectures can be created and used. A detailed literature review is done to determine how programming agents may be done and what kinds of applications agents can be utilized for. Additionally, in recent years, AI has been developed for use in projecting traffic demand. Forecasts of the weather and traffic conditions are used to manage and control traffic, relieve congestion, and make snap decisions in potentially dangerous situations like traffic accidents. Additionally, it can assist in determining whether to build new infrastructure or add lanes, which path to take in the event of an accident or inclement weather, as well as how much money is required for maintenance and repairs. However, automated public transportation systems and autonomous cars are rapidly using AI approaches to prevent delays, mishaps, and traffic jams.

KEYWORDS

- **artificial intelligence**
- **automated vehicles**
- **intelligent agents**
- **intelligent transportation**
- **smart management of traffic**
- **smart mobility**
- **smart network**

REFERENCES

1. Agre, P., & Chapman, D., (1987). PENGI: An implementation of a theory of activity. In: *Proceedings of the Sixth National Conference on Artificial Intelligence (AAAI-87)* (pp. 268–272). Seattle, WA.
2. Sadek, A. W., (2007). Artificial intelligence applications in transportation. *Transportation Research Circular*, 1–7.
3. Allen, J. F., (1984). Towards a general theory of action and time. *Artificial Intelligence, 23*(2), 123–154.
4. Barringer, H., Fisher, M., Gabbay, D., Gough, G., & Owens, R., (1998). Metatem: A frame-work for programming in temporal logic. In: *REX Workshop on Stepwise Refinement of Distributed Systems: Models, Formalisms, Correctness (LNCS)* (Vol. 430, pp. 94–129). Springer-Verlag: Heidelberg, Germany.
5. Bates, J., Loyall, A. B., & Reilly, W. S., (1992). An architecture for action, emotion, and social behavior. In: *European Workshop on Modeling Autonomous Agents in a Multi-Agent World* (pp. 55–68). Springer, Berlin, Heidelberg.
6. Doˇgan, E., & Akgüngör, A. P., (2013). Forecasting highway casualties under the effect of railway development policy in Turkey using artificial neural networks. *Neural Computational Application, 22*, 869–877.
7. Akgüngör, A. P., & Do ˇgan, E., (2009). An artificial intelligent approach to traffic accident estimation: Model development and application. *Transport, 24*, 135–142.
8. Luˇci´c, D. B., & Teodorovi´c, P., (2001). Bee system: Modeling combinatorial optimization transportation engineering problems by swarm intelligence. In:

Proceedings of the Tristan IV Triennial Symposium on Transportation Analysis, Azores, Portugal (pp. 13–19).
9. Farmanbar, M., Parham, K., Arild, A., & Rong, C., (2019). A comprehensive examination of smart grids in the context of smart cities. *Energy, 12*(23), 4484.
10. Masera, M., Bompard, E. F., Profumo, F., & Hadjsaid, N., (2018). Smart (electricity) grids for smart cities: Assessing roles and societal impacts. *Proceedings of the IEEE* (Vol. 106, No. 4, pp. 613–625).
11. Neffati, O. S., Sengan, S., Thangavelu, K. D., Kumar, S. D., Setiawan, R., Elangovan, M., & Velayutham, P., (2021). Migrating from traditional grid to smart grid in smart cities promoted in developing country. *Sustainable Energy Technologies and Assessments, 45*, 101–125.
12. Lai, C. S., Lai, L. L., & Lai, Q. H., (2021). *Smart Grids and Big Data Analytics for Smart Cities.* Springer International Publishing.
13. Berka, C., & Lewindowski, D., (2005). Implementation of a closed-loop real-time EEG-based drowsiness detection system: Effects of feedback alarms on performance in a driving simulator. *International Conference on Augmented Cognition.*
14. Zantalis, F., Koulouras, G., Karabetsos, S., & Kandris, D., (2019). A review of machine learning and the internet of things in the context of smart transportation. *Internet Futures, 11*(4), 94.
15. Gaspar, J. G., Brown, T. L., Schwarz, C. W., Lee, J. D., Kang, J., & Higgins, J. S., (2017). Evaluating driver drowsiness countermeasures. *Traffic Injury Prevention, 18*(sup1), S58–S63.
16. Liu, M., (2021). Smart urban transportation is built on big data. In: *Journal of Physics: Conference Series* (Vol. 1972, No. 1, p. 012092).
17. Mercy, W. (2018). *Smart Driver Monitoring System to Prevent Accidents Using Machine Learning Techniques*. ISSN: 2456-172X , *3*(3), September-November, 112–124.
18. Papageorgiou, M., Diakaki, C., Dinopoulou, V., Kotsialos, A., & Wang, Y., (2003). Review of road traffic control strategies. *Proceedings of the IEEE, 91*(12), 2043–2067.
19. Abdulhai, B., & Pringle, P., (2003). Autonomous multiagent reinforcement learning-5GC urban traffic control. In: *Annual Transportation Research Board Meeting.*
20. Gartner, N., Stamatindius, C., & Tarnoff, P., (1996). Development of advanced traffic signal control strategies for ITS. *Transportation Research Record, 1494*, 98–105.
21. Guberinic, S., Senborn, G., & Lazic, B., (2007). *Optimal Traffic Control: Urban Intersections*. CRC Press.
22. Tefft, B. C. (2014). Prevalence of Motor Vehicle Crashes Involving Drowsy Drivers, United States, 2009–2013 (Technical Report). Washington, D.C.: AAA Foundation for Traffic Safety.
23. Klauer, Sheila Dingus, Thomas Neale, Sudweeks, T., & Jeremy Ramsey, D. (2006). The Impact of Driver Inattention on Near-Crash/Crash Risk: An Analysis Using the 100-Car Naturalistic Driving Study Data, 594.
24. National Sleep Foundation. (2008). National Sleep Foundation: 2009 Sleep in America® Poll – Health and Safety (Version 2) [Dataset]. Cornell University, Ithaca, NY: Roper Center for Public Opinion Research. doi:10.25940/ROPER-31115319.

25. Buehler, Ralph. (2009). Buehler 1 Promoting Public Transportation: A Comparison of Passengers and Policies in Germany and the U.S. Transportation Research Record. 2110.
26. Taimoor Abbas & Fredrik Tufvesson (2013). Line-of-Sight Obstruction Analysis for Vehicle-to-Vehicle Network Simulations in a Two-Lane Highway Scenario, Radio Wave Propagation and Wireless Channel Modeling 2013, *International Journal of Antennas and Propagation,* 1687–5869.

CHAPTER 5

Blockchain for Efficient and Improved Strategy in Power Stations

KULDEEP SINGH KASWAN,[1] JAGJIT SINGH DHATTERWAL,[2] and PREETY[3]

[1]*School of Computing Science and Engineering, Galgotias University, Greater Noida, Uttar Pradesh, India*

[2]*Department of Computer Science and Applications, PDM University, Bahadurgarh, Haryana, India*

[3]*Dr. Akhilesh Das Gupta Institute of Technology and Management, New Delhi, India*

ABSTRACT

The energy sector has historically been reluctant to implement modern technological technology. Current digital development strategy affecting all actors in the energy market would necessitate the transition to a new energy market. While this still appears to be an unlikely and challenging scenario for major players to consider, particularly in the market, pilot programs have demonstrated that, although the framework is still in the early stages, blockchain will play a significant role in the future energy sector process of acceptance. This chapter aims to explain an alternate method of irrigation for smallholder farmers. The alternative is to establish associations of small farmers using renewable energy for irrigating agricultural fields from solar cells. Not only digital technology linked to solar panels and pumping stations can track the performance of the proposed scheme, but modern blockchain technologies can also be used

Smart Mobility and Intelligent Transportation Systems for Commercial and Hazardous Vehicles.
T. Poongodi, Naga Swetha Pasupuleti, and Balamurugan Balusamy (Eds.)

to encourage energy innovativeness and development and improved processing levels through intelligent agreements. A Solar Token variant identical to Bitcoin blockchain, a utility token, which provides new opportunities for electricity and water trafficking, has been proposed to speed up the transition to the renewable energy system.

5.1 INTRODUCTION

Upwards of one-half of European-generated electricity from alternative energy sources is projected to come from the EU in approximately 12 years. Electricity reducing carbon emissions is expected to be completed by 2050 [1]. In these circumstances, the whole energy market will have to undergo a significant transition, affecting everyone from production, use, manufacturers, and networks in the business process [2].

In recent years, solar power has been used in diverse applications and is one of the most practical uses in the irrigation sector. This way, many farmers will increase their benefits by lowering the energy or fuel price [3]. This approach attracts many farmers. In addition, solar cells could still be used with a high potential for supplementing the power required for farm operations during the winter after the irrigation interruption. Solar power models are equipped in remote areas with no connections to the national power infrastructure [4].

The photovoltaic (PV) device is simple to build since the networked systems have only a limited number of components and no need for the room because the yearly energy consumption is understood.

Analysis shows that plantations can also be used with water-dissolved unfiltered water or organic nitrogen fertilizer [5].

The science of solar cells is more straightforward. It is not reliant on extreme sunshine. The innovation is sufficiently easy to incorporate quickly. To make money from cultivation, groundwater is obligatory. Farmers use electricity when pumping groundwater, but power stores market it at an unfairly high premium to the farm owners' earnings. Due to negative motives, current irrigation channels degraded over time due to discontinuation [6].

Renewable energy sources from the use of Solar panels (photovoltaic) panels, who have used these possibilities to water farmland.

The experiment does not preserve the extra resources in the battery. Still, it exchanges the excess in the periods of higher luminosity in cooperative firms that are a part of all the other cooperative societies or farming in the national scheme [7].

Zero higher requirements are incurred concerning maintaining the cleansing panels, and automated cleaning mechanisms for mini-robots exist in the case of a wide range of boards. The potential absence of a national energy grid in the farm's field is situated (cooperative) [8].

Given the implementation of solar energy costs, any alternative revenue stream such as Solar token can be proposed for small breweries. Centered on the bitcoin blockchain, SolarCoin is globally and decentralist, much like Blockchain, but it is far more sustainable and environmentally friendly in many respects [34]. SolarCoin distinguishes itself by having an ecologically and socially responsible contribution to vertically generated Solar Energy instead of other forms of cryptocurrency [9].

The analysis shows that these innovations contributed to the sustainable development objectives by creating a range of improvements and activities amid the challenges of using Blockchains. Blockchain technology was considered for the irrigation method [10].

We suggest a software solution using the technologies of Ethereum blockchain and Distributed Ledger that aims at reducing farmers' costs in connection with irrigation.

Australia has developed and released a very related cryptocurrency computing application – Energy Registry is a platform for corporates and customers to exchange resources peer-to-peer. The initiative mainly provides renewable energy, including solar and wind power, generated, and exchanged by private corporations and individuals [11].

The key goals and new features of this analysis are:

1. Proposition of the latest solution that would ensure a stable and open energy tracking solution by decentralizing information collected and using blockchain technology for the small-producer partnerships, in a manner most advantaged for both consumer and producer [12].
2. Present the potential to acquire power that blockchain coins and Solar token innovations and report presents the results to select the remedy that will have the more significant advantage over association size [13].

3. Recommending the placement of solar power parks to ensure the farmland cannot be influenced by agriculture and to preserve higher quality water deposited on the network [14].
4. Review of the new irrigation system stage used as a case study for implementing the PV system and the proposal of alternatives to address problems previously found.

In addition, this study introduces a PV system energy generation that can be connected, when more irrigation energy is needed, to the proposed photovoltaic water pumping (PWPS) [15].

So, the chapter structure is as follows: various choices of the first advanced methods for energy transaction fees and the SolarCoin version from Blockchain are provided in the second part. Section 5.3 describes the energy flow model for a farmer's organization—the experimental analysis for renewables irrigation systems. Based on the critical findings provided in an appendix, the configuration of the PV energy system required for supplying the association's water pumps and SCADA system and analyzes of energy produced by the PV system for one year are also addressed [16].

5.2 CURRENT IRRIGATION SYSTEMS IN DIFFERENT COUNTRIES

Concerning water balancing in Europe, energy generation accounts for 44% of the water taken; agriculture accounts for 24%, the population for 21%, and manufacturing for 11%. The agriculture industry represents 60% of total water obtained in southern Europe, exceeding 80% in certain regions. This allocation varies based on the area [17].

The adjustment of the precipitation regimes is expected to lead to more freshwater supplies in some areas of Europe in the future, and other parts cannot address the total need.

Irrigation of crops is an environment in which creative methods and current policies will make effective water use even more straightforward. Arid or moderately areas include irrigation in southern European countries such as Greece, Italy, Portugal, Cyprus, Spain, and Southern France [18]. Nearly 80% of the water used in agriculture is used for irrigated agriculture in these areas. A scientific method has been proposed in allowing optimal evaluation of the water and energy used in Spain's irrigation scheme.

In Jordan, Daud, and Mahmoud established another example that long confirmed such a system [19].

The estimated radiation from the sun-generated is 4.5 kWh/m^3/day. The water is collected from a very well. The measurements showed average reliability of the device's safety by over 3%. Related experiments have demonstrated high performance for modern photovoltaic water pumping systems (PWPS). These results obtained were compared.

The case is different in Romania. The countryside is extensive, there are many freshwaters, and Romania had a water supply distributed Almost 3.2 million hectares before 1989, totally ruined by robbed or washed away facilities [20].

The bulk of usable irrigation facilities was discontinued from 1989, and others were not even installed. In May 2009, irrigation infrastructure was run on 5,63,000 hectares of land, of which the Water Availability User's organizations had negotiated 5,53,000 hectares (or 99%).

Romania has a Shortfall of 2.6 million non-registered hectares in the sector of irrigation. In Romania, in 2009, there were no figures on irrigation of 2.6 million hectares of land (today, the area is even more significant). Those figures indicate that 3,00,000 hectares of land are irrigated, but only 1,00,000 hectares are currently covered. The irrigated area of 5,60,000 hectares is stated in additional sources.

It is reasonable and required to leverage investment opportunities to purchase smartphone drip irrigation machines (drip or sprinkle) and financing for electricity supply for solar PV pumps [9–11] for smallholder farmers in an area without electricity supply [21].

Romania has become one of Europe's leading agricultural prospects with a potent hydrogeological potency and essential advantages for organic agriculture (Land emissions from agriculture are also very poor compared with other European countries) [22].

Unsurprisingly, a large portion of the farming region has been abandoned due to the depletion of the ancient water sources and the lack of a clear governmental initiative or another feasible alternative.

High investment is required in rehabilitating water sources. Still, it is possible for small-scale farmers not to purchase water from rehabilitated water resources since the cost of the water supplies is high compared with the revenue earned. Generating confidence in emerging environments is the most significant benefit of distributed ledger technology (DLT) [23].

Considering the safety threats of this form of the framework with a fully integrated autonomous digital infrastructure, potential risks cannot be overlooked. The Ethereum blockchain is the most commonly used in designing and implementing intelligent contracts and is currently used for various decentralized applications. The group is now updated as an open-source platform and one of the best frameworks to execute intelligent contracts [24].

All software packages require simple programming standards such as in-depth protection, checking the units, sanity checks before and after state, code correctness, and daily audit.

The output of a supervisory control and data acquisition (SCADA) supervised photovoltaic panels as illustrated show and of Mobile Online Monitoring system further emphasizes their formulation's potential to agricultural production [25].

As has been seen, agricultural production poses challenges because of water resource management and prices for energy purchases. The work discussed should not evaluate the effectiveness and the connectivity possibility of blockchain technologies used to serve society. The same goes for controlling the agro-food supply chain by using IoT data for energy usage [26].

It is thoroughly defined how to use an irrigated irrigation pumping system. This approach satisfies and quantifies water demands irrigation and redirects energy sources to the essential coverage in the region for PV systems and the economic power excess [27].

Therefore, an observational approach is conducted to the farmers and organizations of the sustainable technology irrigation systems to demonstrate the SCADA interfaces and their core features and benefits in the integrated blockchain framework instead of the above suggestions.

The location chosen for the analysis is located in Oltenia, Romania, a geographic area where the amount of solar radiation is intense and naturally diminished [28].

It is an area of poor financial and social potential. The region being studied is known as agriculture. Growing the Land is most family members' primary livelihood. The current water source scheme states that it has worked in a progressively extensive area of degeneration, partially, year by year: pumps and instruments are destroyed, networks blocked, and pipelines are accused of stealing from the ground indirectly [29].

The mechanism was rehabilitated with the assistance of the International Bank. But, even afterward, it only succeeded partly due to the logistical and nonpractical process. Since 2010 the energy subsidies required for the launch of the pumping have been canceled, which prohibits smallholders from bearing the costs, called State assistance.

This region has a high-energy, efficient solar cell garden. A non-functional irrigation scheme occurs in the proximity canals that are presently not in service. Farmworkers can't use groundwater during drought conditions (spring-summer) for irrigating crops, even though the water is provided for the canal [30].

Although channel restoration and pumping plants are posed, smallholder farmers cannot consider the connections to these facilities as operating water rates in the network are incredibly high relative to their revenue. In this report, three alternative solutions to irrigation will be explored: the capacity to use sustainable energy using PV cells with their constant decrease in cost, their concentrating of micro-networks, and the likelihood of developing their viability from the very outset by controlling the output of the solar cell find parking to the pumping station:

- National electricity source for soil enhancements and water supply to poor farmers from the national energy system;
- The use of electricity produced by a neighboring solar cell plant which supplies the power to the national energy grid or which farmworkers can use for the reconstruction of the current irrigation system in the supply of water pumps;
- Forming an organization to control and manage energy from renewable sources and regulate irrigation facilities networks.

The restoring irrigation method had already checked the first approach in our country with European funds in another agricultural region. Considering the evidence provided in — the Legislative report showing that the irrigation scheme is failing — it is evident that, regarding the pumped situation, individuals are unreliable or ought to be restored to a smaller federal government or more limited degree [31].

The development of distributed power generation relying on alternative sources is possible with electronics, IT, and energy storage solutions products. The high prices and the unpredictable and intermittent existence of some of these energies will only make it difficult to pursue a far-reaching solution in the future. If no substitute and capitalization are present, or in other words, technology can become far more sophisticated and, of course, cheaper than it is now. But, because they're still humanity's energy expectation, the possibility of using other energy sources (unconventional energies) is very nearby. For both these purposes, the latter approach seems to be the most profitable strategy that small producers can consider, and this experiment shows [32].

5.3 SOLAR PANEL ENHANCED ENERGY SECTORS

The solar panel is an alternate cryptographic panel using blockchain with the proof of staking facilitating solar power generation. The methodology enables the energy produced and used to be safely quantified. Suppose the electricity is provided by in a universe. In that case, solar energy is more critical than in the traditionally generated energy sector. The owner of solar cryptocurrency — individuals or broad solar panels — proposes the creation of a currency transaction.

SolarCoin reflects tokens issued during the management of solar power output. They can also be traded as ordinary cryptocurrencies, as they exist on four different public markets for blockchain technology. All transactions in particular blocks are obtained, tracked, and synched to create a SolarCoin blockchain technology. A highly trusted database underpinning this project is a prominent, confident, and verifiable leader of the total power generated by panels for panels. It pays solar generators for 1 SLR per 1 MWh of solar energy produced with cryptocurrency currencies [33].

A POST (stake period concrete evidence) blockchain technology originating from the implementation of the Vericoin construction process is the technology behind it. SolarCoin is a safer alternative to proof-of-working blockchains than Bitcoin since miners need to solve programming puzzles.

SolarCoin is a minimal charging additional incentive that does not focus on another stimulus that the PV producers will obtain (Including subsidized loans, renewable power rates, tax breaks, green certification, etc.). SolarCoin is an incentive to generate solar energy to lower CO_2 emissions.

1. **Economy Solar Production:** The PV holder is eligible as the lowest incentive for solar power production through SolarCoin. The entitlement to bitcoins depends on a purchaser of electricity generated instead of the money holder for power panels with an output of somewhere around 20 kW [34].

2. **Principle of Working:**

   i. **Step 1:** The holders must declare their services with the SolarCoin Foundations to be considered for a SolarCoin award. Authorizations may be made through a SolarCoin affiliate or directly monitoring application on the family foundation websites. It requires information about the owner of the facility, Service provider link (individual verifiable), solar generation (e.g., meters translations), and the KYC data (know-your-customer).

   ii. **Step 2:** If the registration is completed, the Fund immediately creates SolarCoin in the digital wallet of the SolarCoin holders.

   iii. **Step 3:** SolarCoin is distributed to licensed plantation holders every six months.

   The SolarCoin becomes good with growing liquidity and is published and managed transparently by a cryptocurrency, which can be saved offline or online wallets (e.g., paper wallets).

   When the news (e.g., with the explosions in the Amazon making pollution challenges increasingly apparent, renewable energy gains more and more coverage [35]. This assumes that solar energy suppliers would earn an extra income and demand from SolarCoin, and price increases will increase in the years to come. Holders who produce solar power might use/sell the energy to receive, trade, or reuse SolarCoin.

3. **Maximum Production:** SolarCoin may be an effective stimulant for geothermal energy shareholders to maximize production capacity. Venture capitalists will undertake to invest in renewable energy power through:

   i. The installation price;
   ii. The price it produces for the solar energy;
   iii. Planned revenue from any more rewards.

To accomplish a new condition, SolarCoin uses the exceptional features of blockchain technology:

- A zero-charge extra premium for the output of solar power;
- The first wireless natural funds conservation coin;
- First non-governmental, national, and decentralized stimulus initiative for solar energy.

SolarCoin will also award collaborations — both of which further create the ecosystem. SolarCoin guarantees the electronic tokens can be circulated in a profitable and legitimate environmental and economical way: vertical PV-derived electricity. SolarCoin can evolve, independent of any other national or local stimulation, into a great, infinite PV stimulus. It is important to note that any user of cryptocurrency explorers can display all transaction records transparent and visualization them. Blockchain technologies can minimize costs and length of information management to their delivery [36].

More specifically, the technology for blockchain is ready to adjust the value exchange to escape the need for radiative recombination distribution companies and banks.

In this background, the research focuses on the development at low prices of advanced land improvement systems for all farmers to use. Table 5.1 presents vital buying characteristics and, in turn, distributes the energy surplus generated by PV panels in a household.

The approach that producers recognize as the one that can offer them great economic benefits can be opted for. Table 5.1 can help choose an ideal solution, but it is clear that there are many other specifics; the current state of the surrounding water treatment is the location and number of future association members, etc. [37].

TABLE 5.1 Features of Buying Quality Power Technologies

Country Name	Power Consumed
USA	175
China	100
United Kingdom	125
Russia	100
Germany	180
France	152
Japan	165
Italy	145
Canada	155

A comparison with the suggested solution is carried out in the following parts of this analysis with the functionality mentioned above for possible implementation approaches.

5.4 THE ENERGY FLOW MODEL FOR FARMERS

The first step would be a joint energy procurement by the farmer's organ farmers within the discussed sector. The more participatory it is to measure the cost time (because of the rule of big numbers) and thus the simpler a discount with the electricity supplier is to negotiate. At the end of the cycle, the incorporated consumers will analyze the customer profiles. In the following year, the gains made by employers will be spent.

Inside the next step, understanding the energy usage, the electricity generated required by the company can be calculated. It can be bought directly from OPCOM for Romania (the Romanian Electricity and Gas Market Operator). Under such circumstances, the agency can profit by acquiring renewable resources, which lead to autonomous power generation and storing renewable technologies through the classical system [38].

Sunlight streams to the planet's surface have an incredible energy potential of 172 billion GW. This energy supply will ensure that demand is continuously increased. This becomes a highly complex issue regarding the total quantity used from a functional standpoint.

A collection of PV plates, a generator, and a pump composed of PWPS. PWPS. The device can involve electricity storage and charging batteries, depending on the requirements. The engine is picked based on the power and the current the machine gives. The motor has to mount an inverter if an alternating current is used. PWPS are cheaper than battery systems without a storage facility and less maintenance. Using a power-saving container to the PWPS is less cost-effective than having a battery, and it is beneficial to use inertia while irrigation is necessary. If the tank is deeply formed, the water must be taken from the PWPS. Solar thermal and electrical radiation are significant resources for Romania's regions. Solar panels can exceed up to 1,000 W/m^2 even though costs are incurred, primarily due to battery lifetime, cleaning (board), and incremental costs.

The regular fluctuation of the water volume poured is taken into account. A value threshold (at which the pump begins/stops) on the ground of the PV panel of the maximum solar thermal radiative forcing of the total solar power pump is set to 450 W/m^2. It begins at 450 W/m^2, with increased solar irradiance, and water flow rises.

The function followed by the PV park in the analyzed region can be used where 6,000 WH is available for a generator.

The associations will create their PV parks as a judicial entity by purchasing PV cells. This will allow the Association to exchange the energy generated both within the Organization and to the adjacent customers of the Organization. In the solar power area, the benefit achieved arrives.

The advantage is gained essentially by inserting into the SOLARCOIN index, by which both farmers and organizations will earn SolarCoin. These may be obtained and invested back for oil and other commodities. The SolarCoin Foundation will search recordings of solar panels purchased as given in operational status.

A blockchain development system will send out the amount of additional energy produced. The association's primary goal is to incorporate a software framework that tracks each user's efficiency.

In the form of goods, the Organization owns the facilities on which the people on whose land are situated are supplied. In this situation, the panels should be positioned over the channels, and water should be hygienically safeguarded in this canal for maximum ground utilization.

On behalf of the organization, the accumulated power excess will be provided to participants who have energy requirements that currently do not support their supply or the sustainable domestic city (Power wholesaler) (*see* Figure 5.1).

FIGURE 5.1 Proposed diagram sustainable energy supply.

5.5 RENEWABLE ENERGY IRRIGATION SYSTEM

A thermoelectric generator is the cheapest industrial option to connect the water pumping system directly. The system consisted of a PV array attached directly to the DC engine of a centrifugal compressor at a high-level definition. This system is easy because of its simplicity (including batteries and the control unit. While plants were watered in the spring and summer months, the plan was tracked for five months to evaluate the volume and diversion of resources pumping over time and solar irradiation's effect on systems efficiency [39].

1. **Choosing Panels:** For the solar energy scale, the first knowledge is essential to ensure the better functionality of the circulating water supply. The second knowledge should describe the case study to generalize the information collected.

2. **SCADA System for Association Members–Users:** The monitoring device was introduced using the program CitectSCADA from Schneider Electric. A simple GUI enables both farmers and the association to monitor and regulate the actual condition of the irrigation scheme.

   An isolation farm, fitted with a PV module to produce the appropriate energy for supplying the water pump (depth) distribution in the irrigation canal, was examined at the beginning of this process. The device condition can be tracked and managed with this functionality: temperature range, basin-level of water, battery power, and moisture levels. The soil can be tracked and managed.

   The Farm owners' organization can essentially communicate with the network via the smartphone application communicating with the backend, the decentralized database, and the SCADA framework for water and energy tracking and regulation.

   Every participant will be provided with a device to monitor and calculate the amount of energy generated to regulate both the amount of energy created by the Organization and the utilization of every participant. Thus, sensors and an alarm system are supplied to alert users to the moisture status and water plants.

Both energy usage and excess power information are processed in distributed ledger files. These details are available for the association and the community so that all individuals know the phase in which they exist well enough. Thus, anyone will control the device output and fix any losses [40].

3. **Solar Power Generation Interpretation:** The initial step includes using renewable energy options in the immediate vicinity if a farm organization cannot produce adequate energy from its resources. After this process, it will be determined if the company can use the benefit to buy PV panels to generate its self-PV park.

   i. **Step 1**: In the first (1st) point, the organization acquires the energy from the national grid to practice its cooperative operation (the association) before it meets the proposed goals. It is evident from the irrigation system that the manufacture and quality of farmers' products will increase even before income irrigation canal.

      The content benefit gained would be reinvested in the procurement of PV panels next year.

   ii. **Step 2:** A PV park is situated by the irrigation canal. Provided that this solar power park is located in the sampling sites, the energy generated every day and, particularly during spring and early summer times, when farmworkers need assistance to supply pumping could be monitored by the SCADA system that exists in that site. The power can be exchanged from the current solar power parks or the stock market by introducing the cooperative (the farmers' association). By the purchase of power, the organization will save money. The funds will be spent on purchasing solar panels granted by farmers from the community. If solar thermal purchasing is significant, the solar energy parks will be created own. In addition, revenue can be generated from the respective PV owners' site using SolarCoin, simply by registration in a repository.

   iii. **Step 3:** The company currently has a private power generation facility; its capital can be produced, and therefore also the manufacturing costs can benefit greatly. Via the network architecture, the energy surplus is transmitted (sold) to farmers from outside cooperative societies on their requirements or to another organization, particularly electricity produced in months when it is not filled with water. The electricity will be supplied at home expense during the time of year when it is not rinsed.

The electricity supply from battery storage parks with 7.5 MW is predicted on the energy generation estimate and efficiency (30,646 components and 14 electric motors protecting 17 ha near the irrigation channel.

Suppose electricity generation is supplied to the pumps from water supply stations in the highest range of months (May–September). In that case, it can be calculated to fulfill the requirements by evaluating the values reported. An appraisal of values for out-of-season months could also benefit from providing or supplying electricity to the national energy grid for the household expenses of the federation's participants. The Suggested Cost Mitigation Strategy Taking into account the ideas supplied and addressed during the first aspect and importance of irrigation for small-scale farmers systems (which they know is essential), this study offers new approaches for producer revenue while retaining low rates for water purchased from irrigation networks.

Different methods were suggested based on the region selected and current resources for the study. The cost savings will be first under the current approach. This can be accomplished primarily by reducing electricity costs through Renewable energies and the proposed exchange process.

5.6 REDUCTION OF COST

5.6.1 CURRENT SYSTEM

Due to the growing prices of the new solution (energy provider from the national power system), poor farmers considered the solution unsustainable and thus undesirable. Thus, clean forms of energy must be used to reduce power prices.

5.6.2 UTILIZATION OF SOLAR ENERGY

Using electricity from a solar park near the farms is an alternative option. The farm owners' organization can effectively be linked and tracked if the PV park has a SCADA Electricity control and distribution system in real-time.

Examination of the data (see the previous section) showed that the owner of this PV park can use an energy-effective device but is cost-efficient. This approach, however, does not generate farmworkers since the power produced is supplied at the distribution company's expense.

5.6.3 EXCESS POWER CAPACITY

A continually increasing market for power and excess capacity, one of the lucrative small enterprises of the next few years is smaller power output. But it's almost challenging to distribute electricity and make money for smaller energy suppliers, though, because currently, only major producers have exposure to the energy sector. As a result, only limited producers' organizations could enter the market. The local dealer would be required, at a set price, to take over the amount generated. The premium is up to 50% cheaper than the prevailing selling price.

This approach would also be helpful to small farmers but difficult to introduce at this point. This solution will benefit from substantial profit. Therefore, the next part will propose a simple, advantageous approach for the farmers' association.

5.6.4 DEVELOP INFRASTRUCTURE

Poor farmers are allying to award a new canal of irrigation. Without really being injected into the regional power infrastructure and then used, energy generated by PV panels is used indirectly.

It will be necessary to equip irrigation canals with PV cells to profit. In light of bid and demand, the organization will continue to grow the PV parks efficiently in the business economy. The water channel should also be filled with floating PV panels that avoid erosion to prevent the land required for cultivation, and the anchors can be less than 25 years or more.

The PV system on the top of the canal has the following benefits in contrast with the power array on the earth:

- The ground does not shield the panels;
- They are not apparent;
- The water in the channel cools the solar plant and thus prevents overheating of silicon-based panels (every cell has a capacity of up to 200 kW);

- Generates from an ordinary “ground” PV park where panel shift (more effective if the panel will move according to the brightness) between 11 and 30% more energy due to water cooling;
- The floating device is placed on a light carrier which passes through the light, enabling the animals to live in the water;
- Lower evaporation of water;
- Halt’s inflorescences of algae;
- Cold weather decreases solar cell aging and improves operating efficiency, and allows for a longer time of operation;
- Needs a smaller, less water-based cleaning method than the planetary solar energy panel over time.

In the months in which the soil doesn’t need additional water or maybe encouraged to expand the capability to satisfy the demand of potential foreign buyers, the organization would be involved in selling the abundance of energy. The members of the organization would also use some of the electricity.

A blockhouse with a small population will be given, the local generation and usage will be calculated as legitimate and relevant analyzed data. The work proposed in other peer-peer solar energy traded ventures is different but more successful than suggested, as discussed in further subsections.

5.6.5 BLOCKCHAIN ALLIANCE MANAGEMENT

The relevant software approach has been proposed for water and energy management in an alliance, adding value by openness and centralizing the results. The course seeks to reduce the total irrigation costs of smallholders’ societies by leveraging blockchain technologies. It proposes the following parts: a backend program that handles all activities and intelligent transaction costs; a smartphone app, which helps farmers to make money by selling power and water; a shared leader base database, which provides a high degree of openness and can create trust in the organization’s producers’ network—the platform’s specific elements.

1. The backend module—the backend component would play a vital role in the suggested software suite. The SQL and a distributed database repository can store accounts (for instance, big chain

DB) with all the information (associations and representatives of associations), including held, power-generated panels, water-source interfaces, programming applications (APIs) for mobile application, etc. It communicates with its clients, blockchain, and databases in the mobile application. The role of DLT in the decentralization of information implies a high level of accountability and the growth of faith in the infrastructure.

Association administrators may use a web interface to navigate the backend module and set up and customize information, including farm account, privileges, identity verification, etc. The ERC20 utility token—The Structural rigidity scripting language will be built on the Ethereum blockchain. The ticket will have a set public demand and will be used to establish an existing exchange market as the currency within the associations. The tokens will be awarded to association members based on the amount of power provided by PV cells held. Water is also purchased using the ticket from the irrigation canal. Farmers can use the software platform to communicate with intelligent contracts.

2. The smartphone app for managing and tracking the energy generated and the water used by the Agency's representatives. It may also exchange the token that helps households purchase water from an irrigation canal outside the Organization.

   Association participants (farmers) use smartphones to communicate with the service token (buying and selling the ticket, energy, and water). The federation's admins and their appliances, PV cells, and water meters will set the latest components in the device. The SCADA system and intelligent water metering with Ethernet access are used.

5.6.5.1 WORKING PRINCIPLE

A multitude of consumers can use the functional theory effectively.

1. **Step 1:** An organization member can view and sign into the account using the mobile device portal. The new participants are free to build recent reports. There will be a wide range of self-serving functionality.

2. **Step 2:** Farming will fill in PV recognition and purchasing documentation. The network will map the appropriate source of power generated and all other tokens received to date, following initialization of the intelligent power image stabilization and SCADA system. The managers of the association will register fresh landowners.

3. **Step 3:** Participants can use the network to trade cryptocurrencies for groundwater from an irrigation canal or take advantage of the situation for electricity (other users of the platform, a member of the association, or an external user).

4. **Step 4:** Users are given the option to mask their sensitive information. As a part of the general data protection regulation (GDPR), the information in the distributed database data bank can only contain pseudo-anonymized data.

If these phases have been completed, the whole system is automated using the token from the internal economy. This definition can be used with a cluster of partnerships or set up for them. To a higher degree, the blockchain provides even more advantages, so participants can virtually eliminate the expense of electricity distributors.

5.6.5.2 ACCESS TO DATA

The identification of the owners and the credibility of the solar cell purchase records are checked using a hash, and the members of an organization, administration, or other involved household incomes can then access it.

Once the participant is inserted into the database, the details on day-to-day electricity generation, the volume of power available for purchase, everyday use of households within the group, and families wishing to join the organization can also be viewed. The same is true of the water used or used and the number of symbols each participant possesses.

5.6.5.3 SYSTEM AUTOMATION

The system can achieve high automation with the SCADA system and an innovative water metering solution. The non-automated operations are the participants who register and link the equipment mechanically to the irrigation canals.

5.6.5.4 IMPLEMENTATION OF A PROTOTYPE

A token, an ERC 20 backend API, a simple web interface, and account-creating, profiles, and send or receive tokens are built in the implementation model.

1. **Smart Contracts:** For interacting with the Ropsten Ethereum Testnet, the interface to the network was developed using Truffle and web3.js libraries.

   The following diagram scanners see the tokens generated with solidity and used in the different pricing blockchain. As seen below, we use the open data contractual library, Open Zeppelin; this is our integration. "Limited Irrigation System Coin" is the contract created and used for an experiment. The builder is an optional feature designated by the builder keyword and implemented after generating the contract. The title of the token's agreement is the long title to express the token arrangement. In our case, it is "SIST," and this symbol indicates the shortened version by which the agreement is to be called. Decimal places define how a token is divided from 0 (not divisible at all) to 18 (maximum Ethereum decimals). The "contractor agrees" represents the number of tickets to be revalued (created on the first deployment). The intelligent contract library contracts used by OpenZeppelin are as follows: ERC20, ERC20Detailed, and ERC20Burnable, which provide a stable way of executing the fundamental token features, such as the transfer and reception of toks. The moderator of the organization may use the above intelligent contract functions as a safety precaution. The manager is called "setTransferAgent," He may deny that such people use the token. Each time a transaction is performed, these functions are named and verify if a particular account will conduct payments.

2. **Deploy the Contract to Ropsten:** The token is distributed with the Truffe platform for the testnet blockchain.

   i. Create an intelligent arrangement with the following order of the console.
   ii. Contract deployment: truffle launch network.

   The blockchain will use the token for transactions between participants and trades for energy or water until the contract is deployed.

The Remix Ethereum IDE will be used for initial practical and security checks.

3. **The Interface for the Mobile Application:** The user interface suggested for the smartphone device should be sleek, simple to use, and very straightforward.

5.6.6 DISCUSSION

- After performing the algorithm, we could set up alliances, generate farmers' accounts, simulation data about water and energy. The blockchain system used in small farmer organizations provides a simple range of advantages.
- Safe blockchain-based payment and transaction networks.
- Blockchain technology has a high degree of openness, as the number of tokens and all transactions is available to each organization member.
- The DLT would increase the morale of partnerships, as both will see related statistics, such as energy produced and water use.

The next move is to incorporate the next edition of the framework into a small farmers' organization. This aims to collect feedback on the relationships between consumers and smallholder functionalities. In addition, the first checks in the Association will lead to the scaling up of the network and facilitate the Exchanges between many alliances or villages and the viability within the institutions of internal economic development. The residential economy maintains highly effective processes and management even in winter whenever water is available; needs are limited.

The first phase in the digital irrigation business revolution and the solar field will execute the cryptocurrency system in just one mix. The next step is to provide for committee or village-to-villagers transactions until the action is verified and reaches a reasonable degree of acceptance.

5.7 CONCLUSION

For remote regions without connection to the national energy grid, PV devices had the benefit of seeking to suggest the restoration of irrigated agriculture.

The proposed approach creates smallholder farmers' associations using the PV panel irrigation method. The position and form of the PV screen and the use of the latest blockchain technologies regulate the performance of the proposed system. Developing an economic infrastructure within the organization of local producers can immensely gain members.

Firstly, demand determines the token value, providing farmers with the ability to sustain the facilities throughout the winter when all agriculture production is low. Second, non-participants who are involved in purchasing the association's resources or water will use the internal market to get the best deal possible at the moment.

The chapter also considers energy efficiency and chain-power needs and offers creative and sustainable solutions to agro-power challenges, interchange smoothness, and human-machine interface (HMI).

An innovative green energy-driven drip irrigation with SCADA interfaces based on the blockchain technology suggested was used to test the objectives of this study.

The principal outcomes of this analysis are:

- The benefits of the planned PV energy board power innovative green energy irrigation system and a blockchain technology system were demonstrated.
- Expenditure can be retrieved quickly if, by this blockchain application, a great deal of farmers is participating in the same partnership (see electricity output vs. budget for the current year).
- The blockchain functionality tokens were contrasted with SolarCoin innovations.

It is valid from studies that the acquisition by an organization of the agricultural irrigation operation is one of the most potent ways to implement cultivation and support the environment by part or, in some cases, ultimately lowering transaction supply of energy. The sale of the surplus power you get often creates a substantial benefit. It can also be remembered that it can be rehabilitated to allow the unclosed and unwashed channels to work.

KEYWORDS

- **agricultural irrigation**
- **blockchain**
- **cryptocurrency**
- **renewable energy irrigation system**
- **solar energy**
- **solar panels**
- **solar power water pumping systems**
- **supervisory control and data acquisition**

REFERENCES

1. *Agri-Environmental Indicator—Irrigation.* Available online: https://ec.europa.eu/eurostat/statistics-explained/index.php?title=Agri-environmental_indicator_-_irrigation (accessed on 2 November 2023).
2. Wu, J., & Tran, N. K., (2018). Application of blockchain technology in sustainable energy systems: An overview. *Sustainability, 10,* 3067.
3. *Water for Agriculture.* Available online: https://www.eea.europa.eu (accessed on 2 November 2023).
4. Galindo, J., Torok, S., Salguero, F., De Campos, S., Romera, J., & Puig, V., (2017). Optimal management of water and energy in irrigation systems: Application to the Bardenas Canal. *IFAC Pap. Online, 50,* 6613–6618.
5. Kuldeep, S. K., Jagjit, S. D., & Krishan, K., (2021). Blockchain of IoT based earthquake alarming system in smart cities. *Book Entitled: Integration and Implementation of the Internet of Things Through Cloud Computing.* Published in IGI Global, ISBN 13: 9781799869818, ISBN10: 1799869814, EISBN14: 9781799869832.
6. Bizon, N., (2017). Energy harvesting from the partially shaded photovoltaic systems—Chapter 7. In: *Optimization of the Fuel Cell Renewable Hybrid Power Systems* (pp. 285–301). Springer: New York, NY, USA.
7. Shinde, V. B., & Wandre, S. S., (2015). Solar photovoltaic water pumping system for irrigation: A review. *Afr. J. Agric. Res. (AJAR), 10,* 2267–2273.
8. Babkir, A., (2018). Comparative assessment of the feasibility for solar irrigation pumps in Sudan. *Renew. Sustain. Energy Rev., 81,* 413–420.
9. Nallapaneni, M. K., Jayanna, K., & Mallikarjun, P., (2018). Floatovoltaics: Towards improved energy efficiency, land, and water management. *Int. J. Civ. Eng. Technol. (IJCIET), 9,* 1089–1096.

10. Wattana, V., Tharwon, A., & Danupol, H., (2019). When blockchain meets the internet of things: Characteristics, challenges, and business opportunities. *J. Ind. Inf. Integer.*
11. Kuldeep, S. K., Jagjit, S. D., & Santar, P. S., (2021). Blockchain technology for health care. *Book Entitled: Healthcare and Knowledge Management for Society 5.0: Trends, Issues, and Innovations*. Published in CRC Press, ISBN 9781003168638.
12. Hua, H., Qin, Y., Hao, C., & Cao, J., (2019). Optimal energy management strategies for energy internet via deep reinforcement learning approach. *Appl. Energy, 239*, 598–609.
13. Hirsch, A., Parag, I., & Guerrero, J., (2018). Microgrids: A review of technologies, key drivers, and outstanding issues. *Renew. Sustain. Energy Rev., 90*, 402–411.
14. Reka, S., & Dragicevic, T., (2018). Future effectual role of energy delivery: A comprehensive review of internet of things and smart grid. *Renew. Sustain. Energy Rev., 91*, 90–108.
15. Zha, X., Ni, W., Wang, X., Liu, R. P., Guo, Y. J., Niu, X., & Zheng, K., (2018). The impact of link duration on the integrity of distributed mobile networks. *IEEE Trans. Inf. Forensics Secure, 13*, 2240–2255.
16. Li, Z., Bahramiradb, S., Paasob, A., Yan, M., & Shahidehpour, M., (2019). Blockchain for decentralized transactive energy management system in networked microgrids. *Electr. J., 32*, 58–72.
17. Karavas, C. S., Arvanitis, K., & Papadakis, G., (2017). A Game theory approach to multi-agent decentralized energy management of autonomous polygeneration microgrids. *Energies, 10*, 1756.
18. Cioara, T., Anghel, I., Bertoncini, M., Salomie, I., Arnone, D., Mammina, M., Velivassaki, T. H., & Antal, M., (2018). Optimized flexibility management enacting data centers participation in smart demand response programs. *Future Gener. Comput. Syst., 78*, 330–342.
19. Jagjit, S. D., Kuldeep, S. K., & Amit, P., (2022). Implementation and deployment of 5G- drone setups. *Book Entitled: Internet of Drones: Opportunities and Challenges*. In Apple Academic Press (AAP), Canada, Hard ISBN: 9781774639856.
20. Ding, W., Wang, G., Xu, A., & Chen, H., (2018). Research on key technologies and information security issues of energy blockchain. *Zhongguo Dianji Gongcheng Xuebao/Proc. Chin. Soc. Electr. Eng. (CSEE), 38*, 1026–1034.
21. Dorri, A., Kanhere, S. S., Jurdak, R., & Gauravaram, P., (2017). Blockchain for IoT security and privacy: The case study of an intelligent home. *Proceedings of the 2017 IEEE International Conference on Pervasive Computing and Communications Workshops (PerCom Workshops)* (pp. 618–623). Kona, HI, USA.
22. Kiayias, A., Russell, A., David, B., & Oliynykov, R., (2017). Ouroboros: A provably secure proof-of-stake blockchain protocol. In: *Proceedings of the CRYPTO 2017, 37th Annual International Cryptology Conference* (pp. 357–388). Santa Barbara, CA, USA, Springer: New York, NY, USA.
23. Wang, X., Yang, W., Noor, S., Chen, C., Guoc, M., & Van, D. K. H., (2019). Blockchain-based smart contract for energy demand management. *Energy Procedia, 158*, 2719–2724.
24. Hertig, A., (2018). *How Do Ethereum Smart Contracts Work?* Available online: https://www.coindesk.com/information/Ethereum/ (accessed on 2 November 2023).

25. Li, Y., Yang, W., He, P., Chen, C., & Wang, X., (2019). Design and management of a distributed hybrid energy system through smart contract and blockchain. *Appl. Energy*, *248*, 390–405.
26. Hassan, M. U., Rehmani, M. H., & Chen, J., (2019). Privacy preservation in blockchain-based IoT systems: Integration issues, prospects, challenges, and future research directions. *Future Gener. Comput. Syst.*, *97*, 512–529.
27. Makhdoom, I., Abolhasan, M., Abbas, H., & Ni, W., (2019). Blockchain's adoption in IoT: The challenges, and a way forward. *J. Netw. Comput. Appl.*, *125*, 251–279.
28. Jagjit, S. D., Kuldeep, S. K., Vivek, J., & Aanchal, V., (2022). Machine learning and deep learning algorithms for IoD. *In a Book Entitled: Internet of Drones: Opportunities and Challenges*. In Apple Academic Press (AAP), Canada, Publishing date, Hard ISBN: 9781774639856.
29. Liu, N., Yu, X., Wang, C., Li, C., Ma, L., & Lei, J., (2017). An energy sharing model with price-based demand response for microgrids of peer-to-peer prosumers. *IEEE Trans. Power Syst.*, *32*, 3569–3583.
30. Unguru, M. *Blockchain Technology: Opportunities for the Energy Sector.* Available online: https://ideas.repec.org/a/iem/eurinf/v2y2018i1p53-58.html (accessed on 2 November 2023).
31. Wang, T., Zheng, Z., Rehmani, M. H., Yao, S., & Huo, Z., (2019). Privacy preservation in big data from the communication perspective—A survey. *IEEE Commun. Surv. Tutor.*, *21*, 753–778.
32. Viji, P. G., Krishna, P. G., Vivek, M., & Ashwini, R., (2018). A survey on security attacks and challenges in bitcoin. *Int. J. Comput. Eng. Technol. (IJCET)*, *9*, 65–74.
33. Conti, M., Kumar, E. S., Lal, C., & Ruj, S., (2019). A survey on security and privacy issues of bitcoin. *IEEE Commun. Surv. Tutor.*, *20*, 3416–3452.
34. Nakamoto, S., (2008). *Bitcoin: A Peer-to-Peer Electronic Cash System.* Available online: https://bitcoin.org/bitcoin.pdf (accessed on 2 November 2023).
35. Ateniese, G., Magri, B., Venturi, D., & Andrade, E., (2017). Redactable blockchain—Or-rewriting history in bitcoin and friends. In: *Proceedings of the IEEE European Symposium on Security and Privacy (EuroS&P)* (pp. 111–126). Paris, France.
36. Ross, K., (2018). *ACWA Power Adopts Blockchain Currency SolarCoin.* Available online: https://www.renewableenergyworld.com/2018/01/18/acwa-power-adopts-blockchain-currency-solarcoin/#gref (accessed on 2 November 2023).
37. *Blockchain: Unchaining Utilities to the Future.* Available online: https://www.numbeo.com/quality-of-life/rankings_by_country.jsp?title=2020&displayColumn=-1 (accessed on 2 November 2023).
38. Hughes, L., Dwivedi, K. Y., Misra, S. K., Rana, P. N., Raghavan, V., & Akella, V., (2019). Blockchain research, practice, and policy: Applications, benefits, limitations, emerging research themes and research agenda. *Int. J. Inf. Manag.*, *49*, 114–129.
39. Zhao, Y., Peng, K., Xu, B., Liu, Y., Xiong, W., & Han, Y., (2019). Applied engineering programs of energy blockchain in the US. Proceedings of the 10th international conference on applied energy (ICAE2018), Hong Kong, China. *Sci. Direct Energy Procedia*, *158*, 2787–2793.
40. Muzammal, M., Qu, Q., & Nasrudin, B., (2019). Renovating blockchain with distributed databases: An open-source system. *Future Gener. Comput. Syst.*, *90*, 105–117.

CHAPTER 6

Artificial Intelligence-Based Smart Cities

JAGJIT SINGH DHATTERWAL,[1] KULDEEP SINGH KASWAN,[2] and ASHOK YADAV[2]

[1]*Department of Artificial Intelligence and Data Science, Koneru Lakshmaiah Education Foundation, Vaddeswaram, Andhra Pradesh, India*

[2]*School of Computing Science and Engineering, Galgotias University, Greater Noida, Uttar Pradesh, India*

ABSTRACT

Artificial intelligence (AI) technology is a strong notion that, if used correctly, can be a vehicle for successful development that might support long-term transformations to a much more starting point of energy livable framework. With its supervised learning features and functionalities, AI may be used as a tool to enable systems to solve challenges that might alter urban spaces as we have experienced them for generations and contribute to the establishment of a new age, the era of the "smart city." Transportation is one of the critical sectors where AI can make a difference. The use of cognitive transportation solutions in general and automation transportation specifically can effectively enhance connectivity availability and influence urban growth. Despite its machine-oriented nature, this new generation of AI-based transit must be a user-centered innovation that "acknowledges" and "appeases" the individual user, businesses, and humanity world in general. Trust must be created for this shift to succeed, and dangers must be removed. This study makes a practical solution that can be applied by extensively discussing the little-studied intersection of

Smart Mobility and Intelligent Transportation Systems for Commercial and Hazardous Vehicles.
T. Poongodi, Naga Swetha Pasupuleti, and Balamurugan Balusamy (Eds.)

AI, mobility, and the smart city and how this will alter urban destinies. It focuses on critical, intelligent transport proposals such as Ability to connect and Highly Automated, fully independent Private and Remotely operated Aerial Vehicles (PAVs and UAVs), and Manoeuvrability (MaaS), but also therapeutic approaches such as the IoT technology and External Internet, as well as broader evolutions such as Industry 4.0. This study is eventually intended to be a fantastic resource for scholars and urban planners, providing systematic and comprehensive explanations of tomorrow's confusing intelligent transport terminology and describing their individually and collectively functions in supporting the nexus in totality.

6.1 INTRODUCTION

Organizations that can reasoning and organize in the absence of a built-in existing information about their surroundings. AI has the potential to make us safer, wealthy, and better by eliminating our reliance on human labor and greatly enhancing our technical and intellectual advancement [1]. Nowadays, AI has revolutionized our society in many ways, from nearly fully automobiles on the highway to automated systems in our households, and it is likely that AI will continuing to infiltrate almost facets of life, from medical services to teaching, amusement, and protection, for the near term [2]. Transportation is one of the key domains where AI will have the greatest framework influence.

ANN (artificial neural networks), GA (genetic algorithms), SA (swarm optimization), the AIS (artificial immune system), "the Ant colony query optimizer (ACO), bee colony optimization (BCO), and the proposed fuzzy modeling" are some of the AI technologies that are making their way into the transportation area. They AI initiatives may have implications for the vehicles, equipment, driver, or transportation user, and in particularly, how these combines interactively to create a transportation service that encourages user engagement and encourages social communication [3].

Because transportation is by far the most important pillar of a city's operation, progress, and profitability, revolutionizing transportation modifies the notion of the city. The evolution of urban landscape and transportation systems is inextricably linked, as indicated by ideas such as the urbanization rent hypothesis and mind goes, which conceptualize the relationship amongst transportation and urban expansion [4]. It is hard to separate the concept of today's new communities from the eventual layout of their transportation networks. The supply of sustainable mobility to combat traffic congestion, congestion, and destruction of the environment using artificial intelligence (AI) capable of giving smarter, quicker, cleaner, and cheaper methods to get about is the pillar, along with telecommunications, ecological implications of smart city initiatives, as well as the future of urban science [5].

The development of smart city is very far from constrained to that same adoption of electronic and internet technology solutions to cities because it has imaginative, financially viable, strategy, assimilating, understanding, and persons character traits, or, as it is described, a smart city is an interferometric, interrelated, and intellectual town [6]. A smart city uses information and communications technology (ICT) to cut expenses, optimize resource use, boost interaction, and facilitate the emergence' standard of living. Whereas formerly the language of transportation experts and urbanists focused on smart development transportation and smart development, this has now evolved or expanded to include digital infrastructure and intelligent urban transportation that are representative of digital age potential [7]. This is evidenced by the great amount of project governance, city implementing initiatives, and collaboratively and throughout studies. Initiatives that have sprung up all around the world, particularly in the transport sector The concept of smarter towns is encompassing in our study, and due to our sustainable transportation aspect, is principally positioned to look into the framework of government prioritizing for sustainability urban expansion via transportation innovations, which, as acknowledged, is still a mostly unexplored topic. The change to an Automation movement paradigm, like any other political and social transition, raises fundamental considerations about how the transition will be handled and what actions will characterize this transition [8].

This study makes a unique solution can be applied by defining some of the essential transportation elements that are expected to be crucial to the Chatbot smart home of the future. Finally, the chapter is a study vocabulary that defines concepts and their aspects that represent and effect the newly independent, Next week's transportation concept will be interconnected, collaborative, and digitized. More particularly, the document meticulously mentions the connection of AI, public transit, and the sustainable grid by making reference to intervention strategies trying to refer to Integrated and Automated Driving, Remotely operated and Private

Aerobatic Automobiles (UAVs and PAVs), and Manoeuvrability (MaaS), but also the Wireless sensor Networks, Physiological World wide web, and Manufacturers 4.0, which are multiple efforts that may influence mass transit in explicitly or implicitly ways [9].

Automated driving is often seen as a technology that might herald the beginning of a dramatic shift in mobility. CAVs are the pinnacle of AI in transportation; AI techniques and "deep learning" (DL) operations are instruments that might substitute human interaction and "drive" CAVs instead. CAVs are expected to be capable of introducing eco-driving and electricity capabilities, as well as improved privacy and protection requirements. Improved road capacity distribution and current traffic monitoring and replacing commute time by time that may be focused on more profitable work Nonetheless, notwithstanding their enormous fanfare, CAVs continue to raise more unsolved problems than conclusive solutions [10]. UAVs, helicopters, and personal aerial vehicles (PAVs) have individual style as a feasible option to managing difficulties especially in the fields of tracking the movement and transportation observation. Truly automated UAVs and PAVs will be extremely successful in the future for commerce and passenger movement; they are projected to redefine travel behavior as we have understood them for generations by increasing horizontally the metropolitan environment and transportation networks of urban spaces. MaaS, a relatively new transportation idea with just a few partial implementations to date, promises digitized bundles of personalized transportation that will substitute privately held automobiles and optimize the usage and combining of multiple transportation options. MaaS will be powered by strong Intelligent machines which provide complete travel options, ticketing, and pricing, as well as reliable informational offerings suited to each consumer's needs [11]. MaaS can maximize its possibilities if it embraces the usage of CAVs through automotive and journey programs as a complimentary alternative to its public transport aspects. Technology developments related to an Intelligence smart city that might effect (or be affected by) mobility solutions, such as IoT, PI, and Industrial 4.0, are now becoming extremely significant and will be studied herein via a transportation-oriented lens. It's because the internet of things (IoT) will enable extraordinary levels of integration among both users and forms of transportation; the PI will revolutionize container shipping performance improvements through electronic, done by machines, intertwined, and big data platforms; and Industry 4.0, the cornerstone of today's new primary data, is dependent on smart transportation [12].

The chapter will now present the basis for the research method used. A summary of the abovementioned projects (one for each) and a concluding part that acts as a glossary for the topics discussed in the preceding portions This section of the report also includes inferences that brought all of the separate parts of the complex mystery of good city maneuverability together and produce some set of recommendations for city geologists, lawmakers, transportation, and city designers, as well as an initiative for areas for future scholar [13].

6.2 LITERATURE SURVEY

This chapter is comprehensive research assessment of peer-reviewed material written in Journals on the intersection of AI, transportation, and the sustainable city. As a quality indicator, the Database search that backed the technique was restricted to published journals only. The first publication ever documented that explored this study goal was released in 1984, however the great number of literatures generated (>90%) were released during the previous decade; these studies were the more pertinent to our task. Furthermore, our study contains a few references beyond this research that emphasize ancillary, but nonetheless significant, areas of the wide agenda addressed that were missed by the Euler finding algorithm but were thought vital to substantiate a few of the primary claims stated [14].

GenBank was used for four distinct searches. All potential permutations of the study's major parts were investigated. To ensure that the majority of the linkages documented in a wide and interdisciplinary research involving AI, transportation, and "the smart city was discovered and investigated in our research the search" terms "artificially intelligent" and "transport" yielded 280 publications, "transfer" and "smart city" yielded 594 articles, and "machine learning" (ML) and "sustainable grid" yielded 201 papers that met our requirements. A most crucial inquiry of all, combining "AI technology," "transportation," and "sustainable grid," supported reasoning for producing this chapter only 19 publications were identified that explore all three parts and their interrelationships [15]. Only two of them publications had a direct, and only partially direct, relationship with the specific aims of the current effort, while the others were primarily computing- and technology, or focused on extremely isolated areas of the nexus mentioned. One especially intriguing discovery was the accessibility of meeting rooms. Papers as contrasted to published articles; we ended up selecting people 201 presentations for the keywords "machine learning" AND "smart city," while the lot of research conference presentations was almost double that amount [16].

The data analysis for the subject combinations mentioned above, as well as quantitation of the resources found. Because of the overwhelming amount of publication publications discovered, only present the five top magazines (in terms of the number of publications responding to our four important keywords) per pairing. We did not exactly end up reading these magazine publications because our analytic criteria were mainly concerned with integrity and applicability [17].

The bulk of these publications were previously explored in our preliminary and more general search strategy. Again, the majority of them did not correspond to the precise study topic of our work and was instead meaningless, with the vast majority out of these being highly specialized.

Finally, every one of the four authors perused the 1668 trophies of both the journal Washington Reuters, then evaluated more than 500 journal articles of publications which were more meaningful to the field of view of current employment, and eventually engendered an individual detailed review predicated on those papers published as more applicable in doodling the photograph of the AI-centric transport network that will underlies today's new smart city initiatives [18]. The writers compared and synthesized their external review drafts in order to create a coherent "considerably larger" narrative, a method led by the main author. In its finished shape, the document includes allusions to 129 research, all of which were reviewed from page to page. The content was chosen based on the particular topic relevance of the academic results, the host standard evaluation score, and the influence of each publication as determined by the number of times it has been cited Whereas the writers recognize that subjectively criteria were utilized in the choosing of the information used this comprehensive literature procedure was used to reduce bias and increase the variety, flexibility, and quality of the information [19]. The writers, each with their own professional experiences, purpose of research, and ideals, worked as safeguards in place for one another, highlighting crucial topics in the material that an individual contributor could not have conveyed in the same multidisciplinary and colorful manner [20].

The writers evaluated data from an identification of relevant worldwide case study instances and headlines to recognize the successes or failures of very early (or pilot uses of) AI-centric intelligent city transportation techniques to corroborate their bibliographical research. This provides an experimental element to the philosophical and methodological qualities of the work, which may directly influence academicians, city researchers, and others. Legislators, transportation providers, and city designers about certain lessons accessible to them that might promote a beneficial change in thinking guided by AI technologies. A semantically similar experimental technique including this framework was utilized with substantial efficiency in the main author's earlier work, which adds an added layer of legitimacy to the current investigation [21].

The work also goes beyond anticipating the "wins," "benefits," and "positive broader economic implications" of the stated AI-centric mobility projects, developing an in-depth knowledge of the less-often-studied impediments, challenges, and unexpected repercussions of this transitional stage. Change, no matter how beneficial it is, is neither simple nor uncomplicated; strategy planning and management are necessary for ushering in the era of the "city of the future [22]."

6.3 INTELLIGENT AUTONOMOUS VEHICLES (AVS)

The vehicle has changed the environmental and constructed environments of the Planet more so than any previous technology in recorded civilization, transforming communities and economy [23]. Self-Driving Vehicles (CAVs) are the organic direct descendants to internal combustion engines, with vibration and action functionality that do not necessitate any sort of transmission line (driver) or wireless power transfer control, as well as interconnection functional areas that contribute to making them aggressive, community minded, well-informed, and synchronized because they can "speak" with about their surroundings and, ultimately, with anything and anything once the IoT is in location [24]. CAVs will alter connectivity, transportation network, and transportation systems by transferring controllability and operating duties from people to machines equipped with massive AI and wireless networking capabilities. CAVs are expected to become the best model of movement, revolutionizing vehicles and city social progress through the application of AI from what it is they are now [25]. Because CAVs are considerably dissimilar from traditional life form cars, the manufacturing sector will need to modernize by completely revamping its manufacturing processes. Automaker, Baidu, Daihatsu, Volkswagen, Dynamic programming, Didi Chuxing, Ford, Renault, Mahindra, and Mahindra [26]. Intelligence-providing behemoths like Search engine and journey behemoths like Uber, which are not currently regarded as pretty standard auto manufacturers, will drastically change the industry in terms dynamic behavior by had become smaller companies in a more highly customizable electric car market, where ICT will be a differentiating great differentiator. The car sector may potentially be compelled to embrace the new business operations. If smart communities opt to aggressively encourage ecological growth by reducing automobile ownership, sales models that prioritize shared usage over individual investment will emerge [27]. Zealand, Germany, Scandinavia, and China are the leading countries attempting to position oneself at the epicenter of the CAV "explosion" by enacting Missile units' legislative changes and going to invest continuous training and development operations such as drone operators and experiments [28].

CAVs are a perfect illustration of how every revolution delivers a spirit of hope and big advancements. It (CAV) are expected to provide substantial advantages, including increased free time (because the driver did spend the approximate of 6 weeks cruising per year); improved highway security and protection procedures; programs that will help, convenience, and in-vehicle riding encounter; conceivably make it much convenient for decision makers to prioritize passenger vehicle and ride-sharing business strategies; and reduced road traffic delays, climate change destruction, poor air quality, noise commotion, and other nuisances [29]. I am presently able to fully operate. At the same time, there have been some misgivings, which are normally less transmitted and researched, about increased susceptibility to cyber-attacks, computer systems weaknesses; invasion of personal and commute personal information profiteering; responsibility distribution obstacles; increased vehicle utilization from more population numbers and uninhabited cars and trucks; steadily increasing dangerous incident, congested roads, and pollutant emissions during the transformation process "when CAVs will co-exist with simplified AVs, moderately and traditional." These possibilities have been documented in the media, although seldom in a specific source [30]. Table 6.1 summarizes the primary possibilities and obstacles indicating and impacting the "full-scale launch of CAVs and their future" utilization predicated on these analyzes. Notwithstanding technical advances, a plethora of claimed benefits, and millions of kilometers driven autonomously in separate and regulated conditions throughout the world, CAVs may be difficult to incorporate. Currently, there is no public transportation in our cities [31]. This is a lengthy procedure that might take several years to complete. This is due to the fact that the advent of these cars is surrounded by uncertainty surrounding their effects on the car-sharing industry and land-use practices, as well as how the public articulates judgments on advantages and issues. An unqualified, technology-driven implementation of CAVs may clash with integration of environmental goals, neglecting the reality that the transportation system is a multifaceted political and social ecosystem that need cooperative optimization of both mechanical and sociocultural sub-systems. Because software is simply one of numerous instruments in the arsenal of transportation, a techno-fix can become a panacea; thus, the revolutionary of CAVs should be supported by legislative, moral, educational, corporate, and community interactions foundations that outline the defined mode of operation [32].

It is also anticipated that CAVs will have a broader function to serve whether they are, like they should be, meant to be a forceful response with environmentally, economic, and livability advantages; a role that goes above today's typical customers with impeccable service regime. They must keep evolving into an addition to numerous pieces to a much more comprehensive and integrated mass transit system that replaces private property rights of cars (and, in some cases, some of their use) with multidimensional movement parcels that advertise unconventional modes of transportation and have CAVs as an able to share possibility on a 'as-needed' basis only. CAVs must become the conceptual cornerstone of MaaS rather than its alternatives. CAVs will thus need to coexist and develop complementary ties with public transportation rather than contending for capacity and prioritized routes. An optimal approach to construction involves transport-oriented developments (TOD) that integrate CAVs to serve as convenient first and last-mile options and as neighborhood connectors to public transit networks. From a municipal and land use perspective, this strategy aims to prevent a further shift towards car-centric transportation habits by seamlessly integrating autonomous vehicles with public transit. While the merging of these formerly distinct transportation systems may pose challenges, future laws and organizational efforts should focus on resolving conflicts and achieving the best fusion of both modes of transport. CAVs must be healthier than it has ever been [33]. Turning electrified is a possibility, but only if the generates electricity from renewable resources and is therefore non-polluting, as highly recommended for the correct changes and changes model. CAVs must be self-sufficient and networked, but they must also be closely tied to electricity and shared usage. According to a study focusing, the combinatorial impacts of mechanization, connection, electricity, and collaborative usage can greatly boost the advantages of CAVs. Unfortunately, the path to transformation is not simple. As according motivate employees to work harder (2020), there are 10 regions of primary consideration for CAV policy and management that must be applied in order for the transformation to be seamless: innovation, legislative action, financial collapse and work opportunities morality, transport networks and territory use, incorporation, reduce accidents, cyberwarfare, and confidentiality, business models, congested roads and travel behavior, and eventually social acceptance, trust, and consumer preparedness [34]. Even if acceptable solution are supplied to all of the aforementioned important issues, the future automotive fleet will be almost homogeneous no earlier than 2050, and only if CAV prices are relatively cheap.

To present, the ResearchGate web browser contains 623 peer-reviewed academic publications on CAVs (connected and autonomous vehicles). When conferences publications and book reviews are included in the query, the figure grows to 1,688 research outputs, with more than multiple of these outputs produced since 2015. Nevertheless, the number of studies investigating AVs is double the greater. This is even though, notwithstanding being connected with Intelligent Transport System (ITS) for generations, connectedness is a later tendency that gives additional interactive social function for a police cruiser; AVs do not simply assume (or essentially completely remove) interoperability, but instead refer to the driver's designed to function without user involvement. This volume of scholarly publications demonstrates the enormous interest in exploring an undertaking with truly transformational potential [35].

6.4 UNMANNED AERIAL VEHICLES TECHNOLOGY

Autonomous Unmanned Vehicles (UAVs), often aerial vehicle (UAV or aircraft systems) uses, are a new strong intrusion that introduces the revolutions of AI and communication technology to commercial flight and aerospace. With the development of high battery, protracted, and minimal segments and sub-devices, low-cost fuselages, and strong micro-controllers and engines, UAVs have the promise to deliver viable opportunities for enhanced defense, enforcement, and economic entities. UAVs can help with intellectual capacity, electronic eavesdropping, and surveillance aircraft, border police, target tracking and type, mitigate, inflict damage and try to hit, civil enhanced security, police departments implementations, pollution management, observing, and geographic information systems operations, earth observation, aerial cartography, environmental monitoring and weather forecasting, forest fire recognition, collision avoidance, seagoing vessels, accident investigation, case of emergencies disaster response, and thus more. That most of these duties serve as key indicators, expressing, and influencing day-to-day life. A modern city's usefulness and maneuverability [36].

UAVs are typically divided into two types: fixed wings and rotorcraft. Fixed-wing UAVs typically have a greater motion and a heavily loaded loading, so they must ensure continuous upwards movements to maintain aloft, making them unsuitable for water pumping such as close examination; rotary-wing UAVs, on the other hand, have mobile cloud computing problems and service module but can keep moving in any manner and continue to stay on the ground in the air. Before complete integrating of aerial innovations in the smarter metropolitan environment, difficulties including the unavailability of an accompanying captain to view and avoiding other planes and helicopters, as well as the significant diversity in remotely piloted aircraft objectives and capacities, must be overcome. Combining UAVs with green infrastructure will result in a more responsible business conditions and a more tranquil housing conditions if problems related to commercial concerns such as morality and confidentiality, as well as price, are overcome. On the one hand, authorization, law, and marketplace acceptance are appropriately handled, while on the other side, technological concerns such as the right use of remote monitoring, communication systems, software administration, resources learning and deployment, and transmission power are satisfactorily resolved. Each of the primary parties in tackling these difficulties has a specific role to play. Professional authorities are developing regulatory requirements, policy proposals, and operational protocols to ensure a safe use of UAVs; researchers are trying to develop frameworks and implementations to construct, combine, and utilize UAV implementations; and the manufacturing is working to establish and create value-added functionalities to pervasive computing using UAVs [37].

"Personal aerial vehicles (PAVs), manned aerial vehicles (MAVs), or personal aerial transport systems (PATS)" are new way to travel that promises to showcase the benefits of land and air movement in order to reduce urban congestion by utilizing open space in the air. PAVs will be a novel mobility solution effective of creating linkages. By providing incredible levels of quick, on-demand urban infrastructure, it will fill the gap between timetabled commercial aircraft and ground transportation. PAVs will probably be able to service the urban spaces of green infrastructure if better technology and lower pricing are accomplished. PAVs will be able to carry passengers without the need for flying abilities, makes them increasingly practicable for everyone. Construction necessities, such as the multitude of airport and car lots, appropriate trying to fly pedestrian walkways, conversation with some other configurations, and efficiencies with development of self-driving vehicles (PAVs and CAVs can stand to gain and accompany one another and), may permit PAVs to substantially lead to the reduction of increased congestion if properly planned and developed. Nonetheless, despite advances in PAV notion processes and design technological advances, there are indeed major hurdles to overcome in associated with public affirmation, reduce accidents, costly facilities, disaster response, disturbing the peace and unwarranted electronic eavesdropping, visual pollution, and inordinate electronic eavesdropping. Concerns about aviation traffic [38].

UAV uses are restricted, while PAVs have been in the initial phases of development, despite promises from Europe's Volocopter that it would begin testing of a hovering "taxi in Singaporean in 2020 and Uber's intentions to conduct passenger aircraft in the United States by 2023. We have over 40,000 Social science citation research" outcomes produced to far, nearly entirely on computational science elements of the two discoveries. According to the books, both are expected to become into huge game movers in public transport and promoters of the urban planning idea in the future.

6.5 MOBILITY BLOCKCHAIN

Mobility-as-a-service (MaaS) is a strong concept that is still in its early planning stages in terms of overall deployment but has received a lot of attention. In its simplest form, MaaS promises digitized bundles of personalized transporting items that will substitute large private automobiles via an all-in-one smart internet site delivering high quality comprehensive tourist experience, reservation, intelligent pricing, and reliable informational functions. If MaaS works as a true alternative for residential personal vehicles, transforming car use to a rigidly as-needed delivery, it has the possibilities to greatly start reducing the quantity of drivers on the road, resulting in significant lowering in transportation disruptions, poor air quality, sound annoyance, electricity utilization, and vehicles and public stigmatization, as well as economic advantages in based on the congestion protection and injury risk reduction, safety and wellness, social cohesion, and community cohesion. It will also make more room, precious living quarters from vehicle driving and parking and redistributing it to municipal and physical transportation, as well as other more human-centric building performance expenditures [39].

These investigations cover a wide range of MaaS topics, such as consumer experience, sentiments, interaction with public transportation, business models, travel behavior, management, and policy, metropolitan possibilities, and even Cryptocurrency adaptation.

Notwithstanding the ambiguity around MaaS application specifications and the lack of a fully unified and uniform definition of the word, these data demonstrate that interest for MaaS research will not be tapering down in the future. There is currently a schism between Researchers and professionals see MaaS as simply a ride- and motor system organizer, while another argue that MaaS must be much broader. This signifies that provided properly delineated and interpreted, it is still very hard to ascertain how MaaS can directly affect driving behavior and how it can support in morphing the prospect; remarked: "there will be very constrained understanding to words when it comes of the based on behavioral influence of MaaS or its potential value." In this perspective, emphasizes the need of studying not just MaaS world in general, but also its constituent aspects (i.e., multimodal travel coordinators, multiple payment, reservation platforms, actual information, and transportation packages) and their impact on customers as facilitators or impediments.

Whim in Helsinki and Wolverhampton are two early examples of MaaS implementations with varying levels of reach, consistency, and effectiveness. The Hamburg pilot UbiGo, which, notwithstanding its achievement. If CAVs are communicated in this manner, they have the possibilities to bridge the divide among both internal and external public transportation completely, making MaaS appear as a less assertive, provocative, and stringent movement framework that, while not entirely fulfilling its governmental hyperbole of "independent unhindered liberation," makes a justification superstar common ground that does include, but is not centered on, car access. On this assumption, the AI-centric online technologies of MaaS will also have to combine and combine multisensory outings.

Three major obstacles stand in the way of the shift to a MaaS-centric transportation system. To discover the most significant impediment to MaaS achievement, focus on the widespread effect of cultural traditions and the need to assist a shift in approach about stewardship, not just for the particularly tech-savvy digital natives, but for all individuals. As people's opinions regarding purchasing vs financing, borrowing, and exchanging transportation vehicles shift, MaaS may become more mainstream in the near future as a result of sustainability concerns and increased environmentalism. Another issue that MaaS confronts is the difficulty in establishing smooth cooperation between the actors accountable for its implementation and manufacturing processes owing to uncertainty in the business strategies and/or competing interests among the transport suppliers that must collaborate. Formulating and implementing, through research and development projects and trials, must thus solve these collaboration challenges by offering unambiguous answers about the best brand of MaaS, because, as the name implies, guarantees of "effectiveness" are not attainable without governmental engagement. The challenge today that we may face is a scenario in which MaaS "assists" public transportation users and travelers without a driver's license (or, in the case of a CAV-based MaaS, without the driving ability) in developing unsustainably high shopping habits by utilizing door-to-door remedies instead of public transportation. Everything which MaaS intends to correct will be a multimodal shift to car-centric alternatives [40].

6.6 IOT-BASED INTELLIGENT TRANSPORT

While the phrase "internet of things" (IoT) is becoming more popular, there is currently no basic definition or knowledge of what the IoT truly means. For many, the IoT evolved as a network environment that conceptualizes a near prospect in which physical items will be prepared with Arduino microcontroller, transmitter, and receiver for online information, and appropriate addressing scheme that will allow them to interact with one another with users, allowing them becoming a world wide web phenomenon. The convergence of various technological and information systems is the primary enabler of these guidelines. The IoT enables remote monitoring, management, and automation of equipment, as well as the generation of innovative discoveries and powerful analytics from huge quantities of actual statistics. It eventually enables an item to hear, see, listen, comprehend, and transmit all at once. The IoT employs a network cloud-based platform that is interconnected. For value generation, technology, sensing, opportunity to contact, 5G, and big data are used.

The IoT is the technological underpinning of smart buildings and what renders them possible; urban areas require three important elements that the IoT can also start providing: knowledge, connectivity, and experimentation. The IoT will supplement the development of ITS, which is founded on the notion of instrument communications.

ITS supply the hardware component of the IoT with equipment such as RFID tags and commenters, wearable technology, guidance systems, and innovative technologies to gather data about driving conditions, vehicle accidents, infrastructure improvements, or the redeveloping of possibilities in the surroundings where they are positioned.

"Two other key facilitators of the AI–transportation–intelligent city the physical interface (PI) and Industry 4.0" will form a nexus. The PI is a revolutionary commercial freight and distribution concept that aims to improve the economical, ecological, and sociological productivity and reliability of how physical objects are carried, stored, realized, supplied, and utilized throughout the world. It is a worldwide open distribution infrastructure associated with physical, intellectual, and operating interconnectedness via encapsulating, connectors, and conventions that will supersede present analog logistically approaches. The PI will convey freight using dispersed, multifunctional transportation systems in which intermediate facilities combine canisters from many origins to optimize loading on subsequent sections. The PI is a unique and sophisticated idea for the development of logistics management, and it will presumably be built on and facilitated by a smart IoT system. In the setting of a modern city with urbanization features and greater internet use the future city's application digital controllers will receive pure and standardized digitized data from digital businesses, a PI-based solution for freight movement. This information will be an additional vital input for the "city brain's" AI and advanced analytics devices, allowing for easier and optimum outcome. "With online interaction and interconnected PI system parts, such as PI-containers" (small packages, palettes), PI-movers, and PI-hubs, the PI, in conjunction with AI, would also provide ultimate decision to create adaptive city circumstances (stores, warehouses). This new and effective choreography of essential aspects of a logistics network will contribute to the pursuit of an intelligent, ecological, and adaptable city. The PI represents the freight and distribution industry in terms reaction to Industry 4.0.

Industry 4.0, also known as the Fourth Industrial, began as a German strategic drive to pioneer industrialization and has now evolved into a disruptive worldview.

After contextualizing and connecting the many components of the future ML smart city transportation eco-system, this chapter will attempt to establish consistent and comprehensive explanations for all of those. Table 6.1 serves as a one-stop shop for the diverse population of city research experts concerned in this subject. This will, ideally, raise the exposure and accessibility of the still-understudied, at minimum in its whole, nexus of AI-based city.

6.7 CONCLUSION

In current study acknowledges AI's revolutionary capacity in the case of smart cities, and how it may be a framework agent that will revolutionize mobility in unimaginable ways. At the same time, it admits that the modern digital transportation paradigm's computer personality must function inside a reasonable, ecological, and user-centered embedded design that "understands" and "satisfies" the individual user, marketplaces, and community benefit of the entire. The promises of self-driving, linked, shared, and digitized transportation services is insufficient if it does not support activities that lead to increased environment preservation, green purchasing, production benefits, community cohesion, integrating, health, and welfare. People must think that a change may be really advantageous to many, that they might becoming active and involved members in the

TABLE 6.1 AI-based Transportation

Intelligent Transportation Terms	Explanations
Artificial intelligence	AI is the abilities of a computer to learn from its surroundings, apply what it has learned to new situations, and ultimately perform tasks with human-level competence.
Smart city	Sustainable, inventive, up-to-date, cost-efficient, and people-focused urban landscapes are what "smart cities" are all about, and they can only be achieved through the adoption of an integrated brand of autonomous, associated, joint, digital, and cloud-based technologies in their tactical decision-making and operations.
Connected and autonomous vehicles	To be considered a CAV, a vehicle must be able to perceive its physical environment, move about autonomously, find its way around obstacles, and act responsibly without any human intervention, while also having the connectivity features that allow it to be active, supportive, up-to-date, and synchronized.
Unmanned aerial vehicles (UAVs)	Unmanned aerial vehicles, more frequently referred to as "drones," are intelligent aircrafts that can fly without the onboard presence of pilots and may offer resilient air transport solutions for the facility of better military, police, and commercial services.
Personal aerial vehicles (PAVs)	PAVs, or personal air vehicles, are a kind of flying people-mover that operate in the open-air space above cities and provide unparalleled levels of quick, on-demand urban travel.
Mobility-as-a-service (MaaS)	MaaS is a system that provides multimodal packages of personalized mobility to replace privately owned vehicles through the use of an all-in-one digital platform that can provide integrated journey planning, booking, smart tickets, and real-time information services on a subscription or "pay as you go" basis.
Internet of things (IoT)	With the help of unified cloud technologies, sensors, software, and human-machine interfaces, the internet of things (IoT) makes it possible for common items to see, hear, listen to, and analyze streams of massive data, as well as interact with each other and with humans.
Physical internet (PI or π)	The PI is a worldwide concept for eco-friendly, effective multi-modal cargo transportation and logistics that makes best use of digital, automated, networked, and big data tools to enhance the flow of material things and their subsequent storage, distribution, and consumption.
Industry 4.0	Industry 4.0 is a paradigm change that involves the use of cyber-physical operations and the internet of things to computerize, automate, digitalize, and inform industrial systems.

metropolitan renewal eco-system, and also that they can use the possibilities offered by the transition wisely. AI–transportation–smart city connection the key to doing this, as with any socio-technical transformation, is to build trust.

Trust should indeed be created by mass communication, public awareness campaigns, R&D expenditures, methodical looking to try and prototyping, and carefully planned progressive adoption. For any techno-centric transformation to take off and maximize its possibility, risks, and highly toxic side-effects, such as the substantial increase in vehicular traffic through conception of more inhabited and (scary) uninhabited CAV trips endorsed by a flawed MaaS business model that may be principally vehicles, should be completely removed. Intelligence development should include strong socio-centric features as well.

As a result, the migration to the next transportation paradigms that will naturally meet the demands of the knowledge economy should be anticipated to be protracted, diversified, and slow-paced, with high degrees of sophistication and ambiguity. Predicting will not always be correct in such a volatile environment as the financial markets. The effective use of AI necessitates a thorough knowledge of the linkages underlying AI and information on the one extreme, and mass transit system features and factors on the others. Making substantial improvements in an era characterized by the living thing, heat of combustion, large private, completely unrelated, and earth police cruiser and its massive effects on local advancement that has defined how urban areas have been engineered and formed for over a couple of centuries necessitates a business strategy approach that relies on perseverance, flexibility, and adjustability. Changes in transportation provision cannot be expedited unless the accompanying technological, legislative, pedagogical, and moral foundations are all in place and sturdy and developed enough to resist a slew of unforeseen and unexpected difficulties. It should go without saying that CAVs, UAVs, PAVs, MaaS, and their enablers— AI, IoT, and PI are not a solution that will fix all that is problematic with urban transportation and urban planning concepts and practices. Despite their revolutionary potential, these compelling cornerstones of the regarding travel galaxy cannot, on their own, resolve all of the difficulties that built surroundings and transportation networks confront short and long term. Instead, they should be considered as complementing parts of the multifunctional and multidisciplinary jigsaw that is knowledge economy, and, in terms of transportation, they should be combined with TOD, autonomous public transportation, leading travel systems, and congested roads strategies. We must also recognize that the sustainable urban story is not just about transportation, but also includes energy, telecommunications, waste treatment, food production, waste-water treatment plants, transportation, embedded system, public security, commercial, and commercial elements that must be addressed. AI, on the other hand, with its industrial control, Networking, accessibility, confidentiality, data processing, data storage, and analytical functions may once again be used to effect good change in all of the essential components of the urban planning architecture. Finally, in accordance with Komninos, our research demonstrates that the influence of green infrastructure extends well beyond the realm of cities themselves, influencing international standing, environmentalism, environmental warming, inclusiveness, and unemployment.

KEYWORDS

- **aerial vehicles**
- **blockchain**
- **intelligent autonomous vehicles**
- **manned aerial vehicles**
- **private aerobatic automobiles**
- **self-driving vehicles**

REFERENCES

1. Kaplan, A., & Haenlein, M., (2019). Siri, Siri, in my hand: Who's the fairest in the land? On the interpretations, illustrations, and implications of artificial intelligence. *Bus. Horiz.*, *62*, 15–25.
2. Gurkaynak, G., Yilmaz, I., & Haksever, G., (2016). Stifling artificial intelligence: Human perils. *Comput. Law Secur. Rev.*, *32*, 749–758.
3. Everitt, T., & Hutter, M., (2018). Universal artificial intelligence. In: *Foundations of Trusted Autonomy* (pp. 15–46). Springer: Cham, Switzerland.

4. Li, L., Lin, Y. L., Zheng, N. N., Wang, F. Y., Liu, Y., Cao, D., Wang, K., & Huang, W. L., (2018). Artificial intelligence test: A case study of intelligent vehicles. *Artif. Intell. Rev.*, *50*, 441–465.
5. Abduljabbar, R., Dia, H., Liyanage, S., & Bagloee, S. A., (2019). Applications of artificial intelligence in transport: An overview. *Sustainability*, *11*, 189.
6. Miles, J. C., & Walker, A. J., (2006). The potential application of artificial intelligence in transport. *IEEE Proc. Intell. Transp. Syst.*, *153*, 183–198.
7. Knowles, R. D., Ferbrache, F., & Nikitas, A., (2020). Transport's historical, contemporary, and future role in shaping urban development: Re-evaluating transit-oriented development. *Cities*, *99*, 102607.
8. Knowles, R. D., (2006). Transport shaping space: Differential collapse in time-space. *J. Transp. Geogr.*, *14*, 407–425.
9. Alessandrini, A., Campagna, A., Delle, S. P., Filippi, F., & Persia, L., (2015). Automated vehicles and the rethinking of mobility and cities. *Transp. Res. Procedia*, *5*, 145–160.
10. Kuldeep, S. K., Jagjit, S. D., & Krishan, K., (2021). Blockchain of IoT based earthquake alarming system in smart cities. *Book Entitled: Integration and Implementation of the Internet of Things Through Cloud Computing*. Published in IGI Global, ISBN 13: 9781799869818, ISBN10: 1799869814, EISBN14: 9781799869832.
11. Albino, V., Berardi, U., & Dangelico, R. M., (2015). Smart cities: Definitions, dimensions, performance, and initiatives. *J. Urban Technol.*, *22*, 3–21.
12. Giyenko, A., & Im Cho, Y., (2016). Intelligent UAV in smart cities using IoT. In: *Proceedings of the 2016 16th International Conference on Control, Automation and Systems (ICCAS)* (pp. 207–210). Gyeongju, Korea.
13. Lyons, G., (2018). Getting smart about urban mobility–aligning the paradigms of smart and sustainable. *Transp. Res. Part A Policy Pract.*, *115*, 4–14.
14. Ahvenniemi, H., Huovila, A., Pinto-Seppä, I., & Airaksinen, M., (2017). What are the differences between sustainable and smart cities? *Cities*, *60*, 234–245.
15. Komninos, N., Tsarchopoulos, P., & Kakderi, C., (2014). New services design for smart cities: A planning roadmap for user-driven innovation. In: *Proceedings of the 2014 ACM International Workshop on Wireless and Mobile Technologies for Smart Cities* (pp. 29–38). Philadelphia, PA, USA.
16. Kuldeep, S. K., Jagjit, S. D., & Santar, P. S., (2021). Blockchain technology for health care. *Book Entitled: Healthcare and Knowledge Management for Society 5.0: Trends, Issues, and Innovations.* Published in CRC Press, ISBN 9781003168638.
17. Kuldeep, S. K., Jagjit, S. D., & Nitin, K. G., (2021). Smart grid using IoT. *Book Entitled: Integration and Implementation of the Internet of Things Through Cloud Computing.* Published in IGI Global, ISBN 13: 9781799869818, ISBN10: 1799869814, EISBN14: 9781799869832. https://www.igi-global.com/book/integration-implementation-internet-things-through/259316 (accessed on 2 November 2023).
18. Papa, E., & Ferreira, A., (2018). Sustainable accessibility and the implementation of automated vehicles: Identifying critical decisions. *Urban Sci.*, *2*, 5.
19. Thomopoulos, N., & Nikitas, A., (2019). Smart urban mobility futures: Editorial for special issue. *Int. J. Automot. Technol. Manag.*, *19*, 1–9.

20. Nikitas, A., Tchouamou, N. E., & Dani, S., (2019). Examining the myths of connected and autonomous vehicles: Analyzing the pathway to a driverless mobility paradigm. *Int. J. Automot. Technol. Manag.*, *19*, 10–30.
21. Nikitas, A., Kougias, I., Alyavina, E., & Njoya, T. E., (2017). How can autonomous and connected vehicles, electromobility, BRT, hyperloop, shared use mobility and mobility-as-a-service shape transport futures for the context of smart cities? *Urban Sci.*, *1*, 36.
22. Barmpounakis, E. N., Vlahogianni, E. I., & Golias, J. C., (2016). Unmanned aerial aircraft systems for transportation engineering: Current practice and future challenges. *Int. J. Transp. Sci. Technol.*, *5*, 111–122.
23. Jagjit, S. D., Kuldeep, S. K., & Amit, P., (2022). Implementation and deployment of 5G- drone setups. *Book Entitled: Internet of Drones: Opportunities and Challenges*. In Apple Academic Press (AAP), Canada, Hard ISBN: 9781774639856.
24. Hofmann, E., & Rüsch, M., (2017). Industry 4.0 and the current status as well as future prospects on logistics. *Comput. Ind.*, *89*, 23–34.
25. Ingwersen, P., & Serrano-López, A. E., (2018). Smart city research 1990–2016. *Scientometrics*, *117*, 1205–1236.
26. Zear, A., Singh, P. K., & Singh, Y., (2016). Intelligent transport system: A progressive review. *Indian J. Sci. Technol.*, *9*, 1–8.
27. Firnkorn, J., & Müller, M., (2015). Free-floating electric carsharing-fleets in smart cities: The dawning of a post-private car era in urban environments? *Environ. Sci. Policy*, *45*, 30–40.
28. Gandia, R., Antonialli, F., Cavazza, B. H., Neto, A. M., De Lima, D. A., Sugano, J. Y., Nicolai, I., & Zambalde, A. L., (2019). Autonomous vehicles: Scientometric and bibliometric review. *Transp. Rev.*, *39*, 9–28.
29. Lu, N., Cheng, N., Zhang, N., Shen, X., & Mark, J. W., (2014). Connected vehicles: Solutions and challenges. *IEEE Internet Things J.*, *1*, 289–299.
30. Chehri, A., & Mouftah, H. T., (2019). Autonomous vehicles in the sustainable cities, the beginning of a green adventure. *Sustain. Cities Soc.*, *51*, 101751.
31. Jagjit, S. D., Kuldeep, S. K., Vivek, J., & Aanchal, V., (2022). Machine learning and deep learning algorithms for IoD. *In Book Entitled: Internet of Drones: Opportunities and Challenges*. In Apple Academic Press (AAP), Canada, Hard ISBN: 9781774639856.
32. Thomopoulos, N., & Givoni, M., (2015). The autonomous car—A blessing or a curse for the future of low carbon mobility? An exploration of likely vs. desirable outcomes. *Eur. J. Futures Res.*, *3*, 14.
33. Gavanas, N., (2019). Autonomous road vehicles: Challenges for urban planning in European cities. *Urban Sci.*, *3*, 61.
34. De Almeida, C. G. H., Milakis, D., Van, A. B., & Hoogendoorn, R., (2016). Vehicle automation and transport system performance. In: *Handbook on Transport and Urban Planning in the Developed World* (pp. 498–516). Edward Elgar Publishing: Cheltenham, UK.
35. Bansal, P., & Kockelman, K. M., (2018). Are we ready to embrace connected and self-driving vehicles? A case study of Texans. *Transportation*, *45*, 641–675.
36. Milakis, D., (2019). Long-term implications of automated vehicles: An introduction. *Transp. Rev.*, *39*, 1–8.

37. Sochor, J., & Nikitas, A., (2016). Vulnerable users' perceptions of transport technologies. proceedings of the institution of civil engineers. *Urban Des. Plan.*, *169*, 154–162.
38. Kougias, I., Szabó, S., Nikitas, A., & Theodossiou, N., (2019). Sustainable energy modelling of non-interconnected Mediterranean islands. *Renew. Energy*, *133*, 930–940.
39. Kougias, I., Nikitas, A., Thiel, C., & Szabó, S., (2020). Clean energy and transport pathways for islands: A stakeholder analysis using Q method. *Transp. Res. Part D Transp. Environ.*, *78*, 102180.
40. Nikitas, A., (2020). Connected and autonomous vehicles: Priorities for policy and planning. In: Vickerman, R., (ed.), *Encyclopedia of Transportation*. Elsevier: Amsterdam, The Netherlands.

CHAPTER 7

Optimized Machine Learning-Based Pedestrian Detection for Autonomous Vehicles

S. CHIDAMBARANATHAN,[1] VISHWANADHAM MANDALA,[2] M. YUVARAJA,[3] and L. MURALI[3]

[1]*Vice Principal, St. Xavier's College (Autonomous), Palayamkottai, Tamil Nadu, India*

[2]*Enterprise Architect, 1311 Horse Shoe Bend, Greenwood, Indiana, USA*

[3]*Associate Professor, Department of ECE, P.A. College of Engineering and Technology, Pollachi, Tamil Nadu, India*

ABSTRACT

Because of the increase in the number of cars, the fatality ratio has been significantly growing in recent years. Systems that can identify pedestrians early on and aid the driver or take action on their own to avoid accidents may be highly useful in lowering the percentage of pedestrian deaths. Pedestrian detection and collision avoidance systems are critical components in the development of an autonomous vehicle. Pedestrian detection has been a highly hard and intriguing topic of research for intelligent transportation system (ITS) researchers. In this chapter, we suggested a machine learning (ML) methodology for autonomous vehicle pedestrian identification. The proposed ML classifier as Naive Bayes (NB), support vector machine (SVM), and random forest (RF). And also 8-Directional Gray Level Co-occurrence is used for extracting the important features from the input image frames. To improve the detection accuracy of the

Smart Mobility and Intelligent Transportation Systems for Commercial and Hazardous Vehicles.
T. Poongodi, Naga Swetha Pasupuleti, and Balamurugan Balusamy (Eds.)

given classifier, by using hyperparameter optimization scheme as particle swarm optimization (PSO), which is used to optimize the classifier to get better detection rate. In this simulation of testing and training of the classifier model by taking VISOR dataset, which has sufficient image by avoid the under fitting of the classifier. The experimental study is conducted by using python tool and the proposed model performance is obtain by using different parametric metrics, which are deliberated in result section. The proposed SVM classifier achieved the better accuracy of 94.13%, RF achieved the accuracy of 92.85% and Naïve Bayes classifier achieved the classification accuracy of 92.98% respectively.

7.1 INTRODUCTION

Pedestrian Detection is a process of determining presence of a pedestrian and marking the direction of movement in a given video or image data [1]. Locating pedestrians becomes a primary task in many smart city applications like forensic surveillance, home surveillance, hospital surveillance on patients and so on. Pedestrian detection is also widely used in intelligent auxiliary driving, intelligent monitoring, intelligent robot, and many other fields where pedestrian activity analysis is required [2–4]. The increase in demand for the pedestrian detection by real world applications, much research are carried out to attain an efficient mechanism for identification of pedestrians. The development in computer vision technologies and improvements in various algorithms aims at achieving accurate pedestrian detection systems [5]. The main steps involved in the process of pedestrian detection, which is deliberated in Figure 7.1.

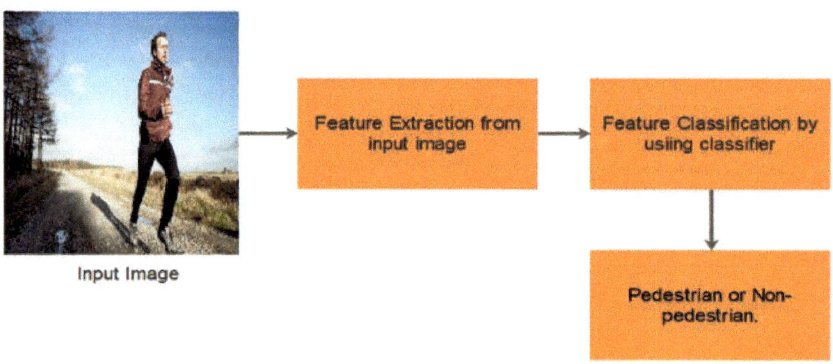

FIGURE 7.1 Basic process of pedestrian detection.

When using a non-stationary camera, detecting moving pedestrians using vision-based methods is difficult. In fact, the ego-motion of the observer causes extra background motion and lighting variations must also be considered. Problems such as these are crucial for methods that detect human movement by removing previous frames. The non-rigid motion of pedestrians makes it challenging to apply optical flow algorithms. Numerous strategies are used to aid in the recognizing procedure. Support the production of hypotheses by combining texture and contour info from the photos with the indication patterns of limb activities. For the final categorization, a temporal exploration of the walking process is used to track the quasi-rigid body portion. For pedestrian detection and tracking, a real-time stereo technique is used. A time delay neural network is then used to classify pedestrians based on their usual leg motion patterns. These methods presume that the walker's leg is visible and restricted to those who are merely walking. Shape-based methods, on the other hand, can identify both moving people and those who are immobile. Most of these systems' challenges come from pedestrian appearance's enormous variability, since they rely heavily on shape factors. Local wavelet features derived from a training instance are used to train an SVM classifier. For this system, the classification accuracy must be sacrificed in favor of processing speed. Once objects are extracted from the background using a stereo-based segmentation approach, a neural network is utilized to recognize them. Short and medium distance detection is possible using stereo vision, and the technology has proven to be reliable.

Initially an image data or an image frame extracted from the video is taken as an input. The image is made up of huge amount of features and the essential features need to be extracted for detecting the input as pedestrian or not [6]. Once the features are extracted the available features need to be classified based on the way the model is trained. Finally, the trained model can be able to classify the input image as pedestrian or non-pedestrian [7]. However, the complexity in real world background, diversified pedestrian postures, diversified shooting angles, demands development of new precise algorithms for pedestrian detection. It is also very challenging to achieve very high detection rates under critical scenes such as in pedestrians at distance. The task of pedestrian detection still faces numerous challenges, necessitating more meticulous planning and optimization. Pedestrian detection can be done using ML for improved computational speed and accuracy, where the traditional approaches suffers from more false detection. There exists various ML algorithms that can be used for detecting the pedestrians. In the existing pedestrian detection approaches the classification of pedestrians are done using filtering techniques [8]. Traditional filters for pedestrian detection are capable to provide good localization, but it depends on scalar valued image features. These filter-based approaches performs well only on the objects with consistent appearance. Advanced filtering techniques works with feature spaces that introduce a specific invariance to small deformations. Classifying the pedestrians using filtering techniques in complex external environments remain challenging [9]. The technological advancements in the area of machine learning (ML) approaches help the system to detect the pedestrians more precisely. In ML approach, the pedestrian detection is carried out in two phases namely feature extraction and pedestrian classification.

7.1.1 APPLICATIONS OF PEDESTRIAN DETECTION

Pedestrian detection has endless applications like person identification, pedestrian tracking and counting, unusual event detection, gender classification, crowd vicinity analysis, fall detection in elderly people, autonomous driving systems, etc. [10]. Person identification Person identification has much attention in intelligent surveillance systems. It is an important task security-based application. The task of application is to identify if the pedestrian in the given the image is present in the gallery of the images collected in the database. This system can be is also used for counting the total number of individual persons present in a given place.

7.1.2 DETECTION OF PEDESTRIANS AROUND AUTOMATED GUIDED VEHICLES (AGVS)

A more challenging application in automated guided vehicles (AGVs) is the detection of pedestrians around the blind spot areas of the trucks [11]. Pedestrian detection in AGVs helps in avoiding the collision of the vehicle on pedestrian, thus reducing the number of road accidents. The goal of this system is to increase the safety of the pedestrians around the vehicle by alarming the driver of their presence, letting the vehicle to slow down or braking on finding the pedestrian nearby, etc. Accuracy in detection is more important in this application to avoid false detection. However, missing of detection could lead to fatal accidents than of false detections.

7.1.3 AUTOMATIC CAPTURING OF WEB LECTURES AND PRESENTATIONS

During the video-capturing process of web lectures and presentations, the pedestrian detection system will automatically follow the lecturer by tilting cameras without the involvement of cameraman [12]. The challenge in this application is the presence of occlusion, where pedestrian detection needs to use multiple models for identifying the human parts separately and then combine for detection.

7.1.4 FALL DETECTION IN ELDERLY PEOPLE

Pedestrian detection can be used in assisting the caretakers for monitoring the activities of elderly persons from remote. This system involves the pedestrian detection and tracking the activity of the elderly person through surveillance camera. If any anomaly activity is detected like fall happened, an alarm or alert can be given to their caretakers [13]. Activity detection remains complex and challenging in this system.

7.1.5 CHALLENGES IN PEDESTRIAN DETECTION

Research on pedestrian detection has grown in popularity for the reason that of the wide range of applications it has, particularly in robots, automobiles, and security [14]. Despite significant advancements, pedestrian identification remains a difficult problem that necessitates the development of increasingly precise algorithms.

7.1.6 CONTRIBUTION OF THIS RESEARCH CHAPTER

- To enhancing the efficiency in extracting the image features by using 8-directional gray level co-occurrence feature extraction.
- Improving the overall performance in classifying the image features.
- To overcome the limitations of ML in working with high dimensional data.
- To identify the best machine model for pedestrian detection.
- To accurately classify the pedestrian for determining presence of a pedestrian in the image.

- To develop a optimize ML model that enhances the accuracy in pedestrian detection.
- To optimize hyper-parameters of machine model by using PSO optimization for enhancing the accuracy in pedestrian detection to minimize the accident in autonomous vehicle.

7.2 LITERATURE SURVEY

Recent advancements in computer vision has enabled active research on detection of humans in sequence of images. In recently, the computer vision community has shown a great deal of interest in methods for detecting pedestrians. On the feature, model and architecture front, pedestrian detection has seen a slew of approaches presented. This chapter focuses on different methods used for pedestrian detections. The study is decomposed into techniques available for feature extraction and classification in traditional, ML and DL approaches. This section discusses the algorithms used in traditional, ML and DL approaches for pedestrian detection and challenges of the existing pedestrian detection systems. The scope of literature is to create new models to improve the performance of pedestrian detection is also discussed.

Hailong Li et al. [15] discussed the challenges existing in pedestrian detection due the complexity of real world background, diversification of shooting angle, and posture diversities. Yi Wang et al. [16] elaborated the difficulty in detecting humans as they take different poses, varied appearance from different shooting viewpoints. They may also appear occluded with other objects or humans. Also included possibilities of false recognitions due to presence of background objects. That has humanoid shapes such as chairs, cylinders, textured areas, or fire hydrant. As these objects are associated with pedestrian shapes and so there are possibilities of false detecting them as humans.

Jong Hyun Kim et al. [17] has discussed the visibility of images at night and with poor lights. Images at night-time usually has advanced proportion of low intensity pixels. There by they have very low contrast. As they do not receive adequate light, they introduce lot of noise. This noise can be reduced significantly if the camera receives adequate light from using some exposure level. Also, the nighttime images introduces motion blurring as the object moves and so it is makes pedestrians to be hardly detected. Increasing exposure at higher level is inadequate in several applications. Even if the experience level is preserved, it had been problematic to extract expressive human features as it would be mixed up with true edges produced by the actual image contents and false edges produced by the noise.

Llorca et al. [18] has compares vision-based pedestrian detection systems in impressive studies. In addition, an analysis of the depth estimate inaccuracy of a pedestrian detection sensor based on stereo vision will be presented. SVM classification was used in conjunction with 3D clustering approaches to detect pedestrians. The results showed that despite the sensor's internal accuracy, it gives adequate measurements due to quantization error. Dollar & Zitnick [19] used random forest (RF) for object detection as two-dimensional representation of bounding box. The model suffered from background clutter and fore ground occlusions. Also becomes insane when data is large scale and complex patterns. Uriel & Victor [20] proposed GA for hand posture recognition. Feature selection was performed by GA and classification by AdaBoost algorithm. The model proved to achieved well but suffers from issue due to local minima.

Zhang et al. [21] have proposed first CNN network to overawed the problematic of cross-scene pedestrian counting by obtaining a density map from the mass image. The model is well trained on a convinced scene and its perspective information, thus achieved robustness to crowd images with scale and perspective distortion. The applications based on this model faces difficulty in acquiring perspective information. Walach & Wolf [22] formulated a layered boosting approach, based on gradient boosting machine and selective sampling. The model proved improvement in the results of regression on pedestrian density map. CNN layers were added iteratively to train each layer for estimating the residual error of the earlier forecast. The method is found to reduce pedestrian counting error by 20–30%, but the accuracy on pedestrian detection count is lower when compared to other detection methods. Also, it could not deal with perspective distortion and cross scene.

Tong Xiao et al. [23] proposed a new DL framework for pedestrian search that involves both pedestrian detection and re-identification. The framework used single convolutional neural network (CNN) that used proposal net for producing bounding box around the target identified pedestrian. The model used input from an image gallery and lacks scalability in the system. Jaderberg et al. [24] proposed VGG16 a pre-trained CNN architecture for object classification. VGG-16 consisting fully connected (FC) 16 layers. The filter are 3 x 3 in size at each convolution layers and stride of 2 x 2 at max pool layers. VGG-16 found to perform well on large, complex dataset. Tome et al. [25] introduced Google Net used for pedestrian detection. They performed a series of convolutions at different scales of filter sizes as *X* 1, 2 *X* 2, and *X*3 with the same padding value 2 X 2 subsequently aggregated the results. The computation was found to be expensive.

7.3 PROPOSED SYSTEM

The system design and architecture of the proposed work for the process of pedestrian detection is represented in Figure 7.2. A video data is taken as an input for pedestrian detection. A video is a collection of image frames, and the image frames are extracted from the video file. These images were given as an input to the proposed pedestrian detection system. These extracted images are given as an input to the proposed system. An image is a bunch of pixels that represents a complete image with collection of objects. Detecting a pedestrian from an image involves extracting the feature set specific to a human from the entire set of available features. Eight-directional gray level co-occurrence is a significant feature extractor method proposed to extract relevant information of the pedestrian from an image to describe it in a more meaningful way. The extracted features were given as an input to the ML model for pedestrian classification. Before we optimize the ML classifier by using PSO optimization technique. Since the three-ML classifier as SVM, RF, and NB scales relatively well to high dimensional data with large number of features over other ML algorithms, these classifiers can be able to classify the pedestrian and not from the input data.

7.3.1 DATASET DETAILS

The proposed work is assessed using benchmarking video datasets namely VISOR dataset (Figure 7.3). These datasets were used for model building and model validation. A total of 8,878 image frames were extracted from VISOR data repository. From the entire image dataset, 5,113 images frames were used for training the ideal and 3,765 images for testing the ideal. The dataset description is labeled in Table 7.1.

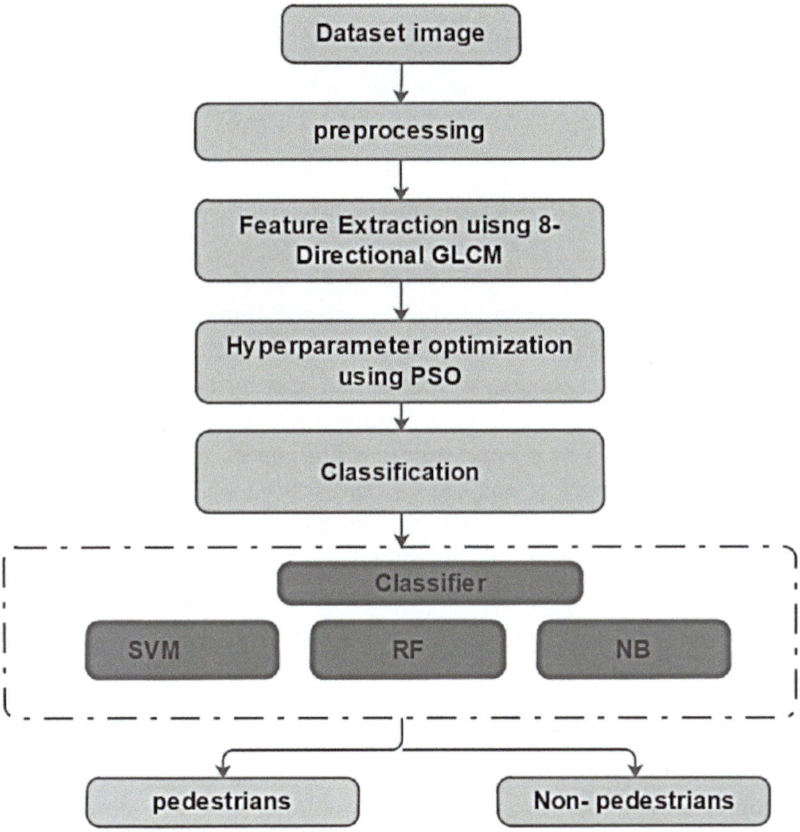

FIGURE 7.2 Proposed methodology.

FIGURE 7.3 Dataset sample images.

Source: https://aimagelab.ing.unimore.it/visor/.

TABLE 7.1 Dataset Description

Total Images	Training – 5113		Testing – 3765	
	Pedestrian	Non-Pedestrian	Pedestrian	Non-Pedestrian
8,878	2,702	2,411	2,004	1,761

7.3.2 PRELIMINARY PROCESSING

An area of interest is defined as the core part of the image where pedestrians are extra likely to be identified and the finding is more relevant for developing avoidance tactics. Using a Sobel operator, the modulus and phase of an object's edges are extracted, and by taking into account the phase of the edges, two distinct binary maps for vertical and horizontal edges are produced. It is necessary to eliminate items in the background at this point before performing the vertical symmetry analysis on the edges. An immovable offset (which can be found through system calibration) can be used to shift either of the two stereo perspectives of a scene, resulting in overlapping of objects that are so far away from the cameras that they can be deemed endless. As a result, background objects can be removed from the vertical edges map using bitwise AND logic and a binary mask. Using a positive threshold, this mask creates a signed difference among the left image and a suitably shifted version of the right image, so the backdrop is black and nearby objects are white. Due to the assumption that far, objects are in the upper half of the image, it is only computed there. It is then augmented using a morphological operator that removes small sized features while amplifying denser regions. In order to enhance groups of pixels while eliminating single pixels, the vertical edges map is first treated with an operator that removes edges derived from detached objects and contextual patterns: Marking specific places with vertical segments of a predetermined length enforces vertical neighborhood.

7.4 FEATURE EXTRACTION

Feature extraction a model of dimensional reductions, maps data from higher dimensions into fewer dimensions retaining the meaning of the data. Feature extraction may be linear like PCA, or Non-Linear like Local Linear Embedding or stochastic neighborhood embedding. While feature selection picks only a subset of the original feature set, feature extraction generates a new feature set by combining various representations of the original feature set, which is the primary distinction between the two processes.

7.4.1 EIGHT-DIRECTIONAL GRAY LEVEL CO-OCCURRENCE FEATURE EXTRACTION

One type of feature is an application-specific piece of information that is useful in addressing computational problems. Structures such as points, edges, or objects may be considered features if they are visible in the photographs. Pattern recognition and ML require feature extraction as a critical processing step. Original data are transformed into a dataset with a smaller number of variables, and this is where the discriminatory information is found. A pedestrian's most basic feature is the ability to change the texture of their environment. When it comes to real-world photos, texture is critical because it describes the substance and structural organization of the surface, such as leaves, bricks, and fabric, among other things. It also describes how the surface interacts with its surroundings. In a nutshell, it's a characteristic that describes a surface's unique physical composition. Texture feature extraction is critical in a variety of image processing applications, including remote sensing, medical imaging, and CBIR, due to the importance of texture information. It reflects the frequency of seeing pixel pairs with various gray levels given the distances between them. Texture characteristics can be generated using this technique. In this co-occurrence matrix, the gray levels are distributed spatially and depend on one another in a particular area of the world. Sets of statistical measurements, referred to as features, can be computed using these matrices.

An image's energy, correlation, contrast, and homogeneity can be extracted using GLCM's statistical method of analysis. GLCM determines whether or not there are any pixel pairs that point in the same general direction of travel [26]. The proposed 8D-GLCM method collects the spatial relationship of gray values with all 8 neighborhood pixels, hence no information is lost. It extracts energy, correlation, contrast, and homogeneity. GLCM can be made in diverse angles as exposed in Figure 7.4(a). GLCM (G [i,j]) built in 8D 45 angles with pixel distance of one has been used in this work. In this work, θ is quantized in eight directions (e.g., 00, 450, 900, 1,350, 1,800, 2,250, 2,700, 3,150).

The steps involved in construction of 8D-GLCM are as follows:

1. **Step 1:** Consider a 4×4 Image I with 4 different gray levels as shown in Figure 7.4(b).
2. **Step 2:** Measurement of GLCM is done with 8 angles and pixel distance of one as shown in Figure 7.4(c).
3. **Step 3:** Computation of 8D-GLCM (*G i,*) The size of the matrix is determined by the formulae:

$$G i, j = Max_{Inten} - Min_{Inten} + 1 \tag{1}$$

where; Max_{Inten} is the maximum intensity value in the image; and Min_{Inten} is the minimum intensity value in the image.

G[*i,j*] matrix calculated from 8 directions (00, 450, 900, 1,350, 1,800, 2,250, 2,700, 3,150) is shown in Figure 7.5.

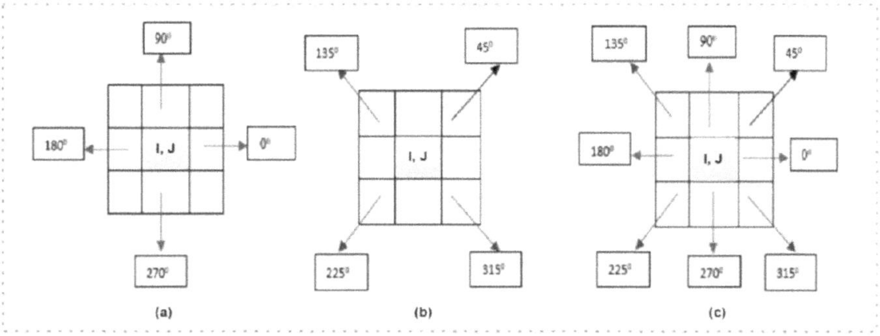

FIGURE 7.4 (a) GLCM construction with four angles (horizontal and vertical); (b) GLCM construction with 4 angles (diagonal); and (c) GLCM construction with eight angles (horizontal, vertical, and diagonal).

4. **Step 4:** Feature Extraction from 8D-GLCM. The 8D-GLCM generates a 32-dimensional feature vector (4 features extracted from each of the eight GLCM leads to 4×8=32 features). From the constructed 8D-GLCM matrix, the following texture features are extracted.

   i. **Contrast:** The local contrast of a picture is calculated using the contrast function. Because of this, the value is low when pixels have similar gray levels.

$$Contrast = \sum_{i=0}^{Ng} \sum_{j=0}^{Ng} (i-j)^2 \; G[i,j] \tag{2}$$

ii. Correlation: The correlation between two pixels in an image is measured through correlation. Uniform distribution is found in textures with a high correlation, while non-uniform distribution is found in textures with low correlation. The following is a list of correlations:

$$Correlation = \sum_{i,j=0}^{Ng} \frac{(i-\mu i)(j-\mu j)G[i,j]}{\sigma_i \sigma_j} \tag{3}$$

where; μ_i, μ_j as follows: (being an estimate of all pixels' intensity in all relationships that contributed to the GLCM) σ_i σ_j (being variance of all reference pixels' intensity in all relationships that contributed to GLCM), N g (being GLCM mean).

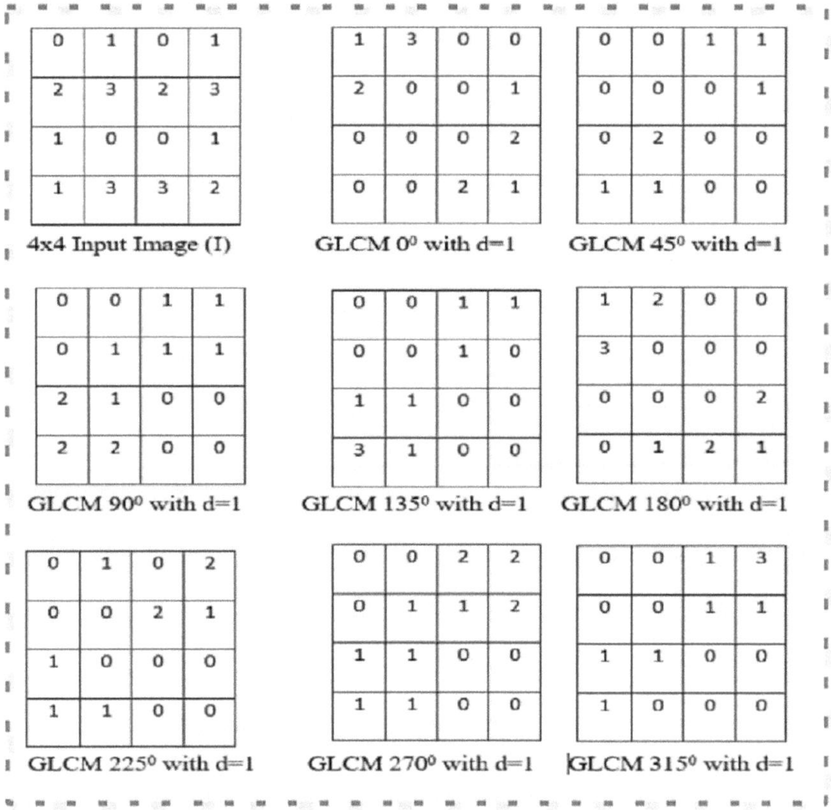

FIGURE 7.5 G[i,j] matrix calculated from eight directions.

7.5 HYPER PARAMETER

ML algorithm optimization relies heavily on hyper-parameters. Hyper-parameters are critical since they directly affect the training algorithm's behavior and the model's performance. The effectiveness of a ML classifier depends heavily on selecting the right hyper-parameters. The model may overlook key patterns in the data if the learning rate is set too little, or it may have collisions if the learning rate is set too great. Hyper-parameter optimization is the process of identifying the hyper-parameters with the greatest potential for improving overall performance.

7.5.1 PARTICLE SWARM OPTIMIZATION (PSO)

A key inspiration for PSO was the way in which organisms behave collectively in nature. A swarm of particles engages in mutual interaction. The position of each particle changes dynamically as a function of velocity, its own best solution, and the best solution found nearby. The initialization of each particle's position and velocity in the swarm is done at random. A correct position shift called velocity is used to update a particle's position. Researchers have developed and tried a variety of approaches for initializing or configuring the particle swarm from scratch. It is possible for PSO to explore the search space quickly and effectively with adequate swarm startup. For initializing particle positions in swarms, uniform distribution along each dimension of the problem space is the most popular approach.

7.5.2 PSO-BASED HYPER-PARAMETER SELECTION

PSO optimizes hyperparameters by looking at two crucial factors: first, how to encode the hyperparameters as the particle's position. Another problem is determining how to define the fitness function, which measures a particle's overall goodness. The following are the two most important aspects:

i. Machine learning classifier optimized hyper-parameters comprise the kernel parameter(s) and regularization parameter (apart from the linear kernel).
ii. Hyper-parameter collection by the PSO requires each particle to signify one possible solution, which is the combination of hyper-parameters. In other words, let's call a set of m hyper-parameters a vector of dimension m. It's possible to choose between two different

kinds of scaling functions as a kernel function, in which case vector is denoted by *n*ð¼*g*;*s*1;*s*2; ... ;*sNinput* Þ *and* v ¼ (*g*,*s*,*d*), , correspondingly, where g is the regularization parameter, d and *sI*(*i* ¼1,2,*y*,*Ninput*) are kernel parameters, Ninput is the dimension number of input patterns.

In this chapter, *v* ¼(*log g*,*log s*1,*log s*2,*y*,*logsNinput*) *and* v ¼ (*log g*,*logs*,*log d*) is for the reason that it provides a more reliable optimization. The parameter vector's length varies depending on the kernel. The fitness function is a measure of overall performance that can be generalized. The generalization performance measure has a number of alternative definitions. As a result, the fitness level that corresponds to that score can be calculated. The following expression is used to determine a particle's fitness level.

For hyper-parameter combination evaluation, the fitness value f_i is used to assess the goodness of particle I and Ave_{test-5} will be employed as in PSO. A good fitness score should show how well a ML classifier generalizes across diverse hyper-parameter combinations. Each particle's average correct rate is used to calculate its fitness value, which is denoted as Ave_{test-5} A PSO-based technique for selecting hyper-parameters in a ML classifier is depicted in Figure 7.6.

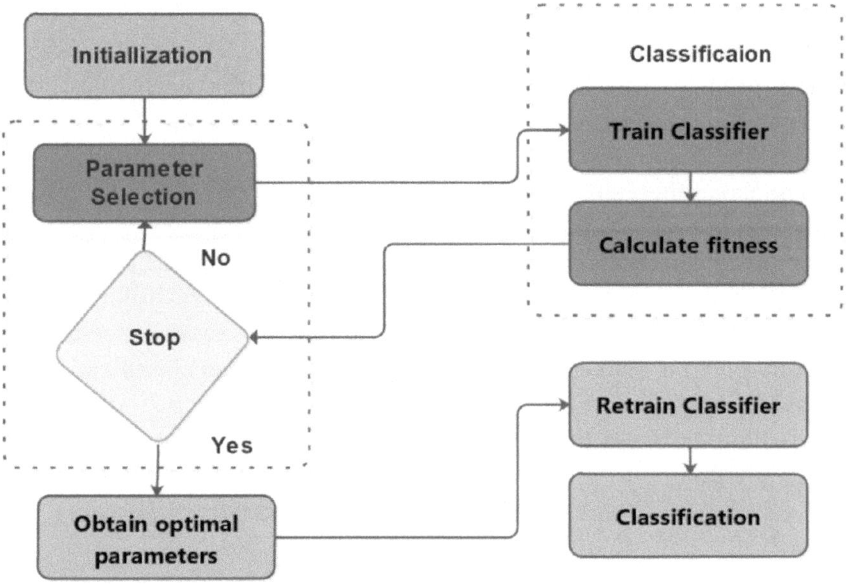

FIGURE 7.6 Flowchart of PSO-based hyper-parameter system.

7.6 NAÏVE BAYES FOR PEDESTRIAN CLASSIFICATION

Naive Bayes (NB) is a ML-based classification technique that produces a probability-based classifier based on the Bayes Theorem. The NB classifier is based on the basic premise that the influence of an attribute value on a particular class is independent of the values of the other qualities [28]. This assumption is known as class conditional individuality, and it allows the NB classifier to be trained quicker. NB uses the Bayes rule, assuming that all accessible vectors in the vector space are independent. The working principle of Bayes' theorem is as shadows. Let X be a data tuple containing a set of features describing the data in a data set. In Bayesian terminology, X H expresses a hypothesis asserting that the data tuple X belongs to a given class C. Tuples representing various classes will have distinct hypotheses. To solve classification issues, it is necessary to comprehend two separate probabilities: posterior probability and prior probability. P(H|X) and P(X|H) are posterior probabilities. P(H) denotes the prior probability. When compared to the prior probability, the posterior probability is based on additional information. Eqn. (4) represents the Bayes theorem and associated probability terminology.

$$P(H|X) = \frac{P(X \mid H)P(H)}{P(X)} \tag{4}$$

where; P(X|H) posterior probability of X, P(X) prior probability of the evidence X conditioned on H, P(H) prior probability of the hypothesis H; and P(H|X) posterior probability of H conditioned on X.

The Bayes theorem in object detection defines the likelihood of being an object based on the circumstances stated as posterior and prior probability that may be connected to that particular item. Applying NB algorithm for pedestrian classification is explained as a process of classifying the input image as pedestrian or non-pedestrian. The variables represented in Bayes theorem given in Eqn. (5) can be applied for pedestrian classification. The conditional probability for a given input object to be as a pedestrian can be represented by Bayes theorem as given is Eqn. (5).

$$P(pedestrain|object) = \frac{P(object|pedestrain)P(pedestrain)}{P(object)} \tag{5}$$

where; P (object | pedestrian) represents the probability of pedestrian images available among the total number of available objects; P (pedestrian) represents the prior probability of total sum of pedestrians; and P (object) represents the prior probability of total sum of objects.

In order to regulate the probability of a specific object to fall into the category of pedestrian or non-pedestrian, Eqn. (5) has to be rewritten in a similar way for non-pedestrian images.

7.7 RANDOM FOREST (RF)

Most of the systems in intrusion detection follow rule-based approaches. These methods like encoding rules had high time complexity and were highly dependent on the prior knowledge of information. Random forests (RFs) suppress this dependency by dynamically changing the model, adopting to the changes in the training sampling. RFs are supervised ensemble learning technique which can be used in classification difficulties. It is more like a collection of decision trees working together which handles well the problem of over fitting in the training phase. These arbitrary sets of simple trees, classify the new instance, where the simple trees vote for instance, and accordingly it is branched into the perfect class. Some of the classic properties of RF are as follows:

- The amount of samples in the training set is N, where N is random, with replacement policy, will be used to grow the tree.
- If α is the total amount of input features, and β is randomly selected, which is less than α for splitting the node. The value of β is kept constant till the forest is completely grown. • Mostly there is no pruning in random forest.
- The correlation among the trees increases the forest error rate.
- If a tree has decreased error rate, then it is considered to be a good classifier, lowering the error rates of all the trees in the forest and decreasing the overall error rate of the forest.
- The choice of selecting β plays a greater role in random forest, increasing or decreasing the value of β will simultaneously affect both correlation and strength. So, more effort should be taken to find optimal β value.

7.8 SUPPORT VECTOR MACHINE (SVM)

Support vector machine (SVM) is considered as a supervised ML procedure that can be used for pedestrian detection. Generally, in computer vision-based techniques, the images are viewed as matrix of pixels, a non-linear representation of data points. The features of target object from the image is extracted and then classified. The features from the images can be extracted, using any of the traditional feature extraction techniques. SVM is a binary classifier that can be used for categorizing the image as pedestrian or not. The general working principle of SVM is the separation of data points into either one of the binary classes, by defining a hyper plane between the non-linear data points of the images. Maximum the distance among the hyper plane and the data point, maximum the accuracy of classification. SVM can also be used for selecting healthy features from the available data points of the images. The process of pedestrian detection in general involves extraction and classification of features from an image or video (Figure 7.7).

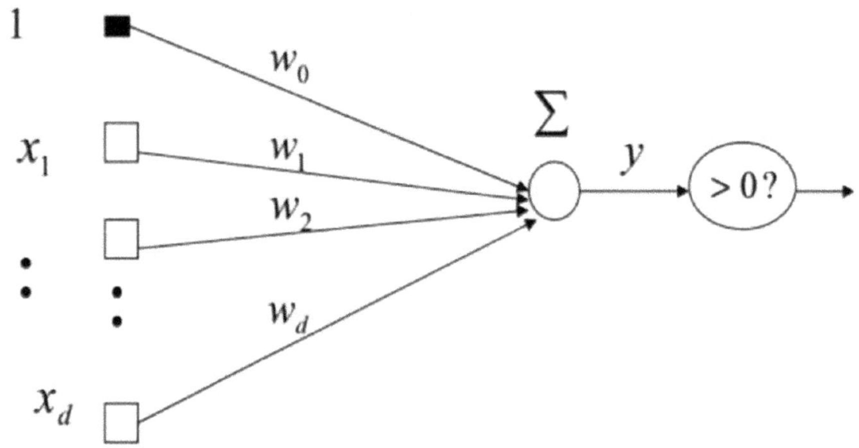

FIGURE 7.7 SVM working model.

where; x_1, x_2, and x_d in Figure 7.7 indicate the picture feature points, while 'y' in Figure 7.7 represent the image class label. When it comes to learning linear or non-linear decision boundaries, SVMs, which are generally two-class classifiers, have been proved to be an appealing and methodical technique.

An SVM algorithm analyzes a set of points to locate a hyper-plane that leaves the most points from one class on either side while also increasing the distance from that hyper-plane from the other. This is the same as conducting structural risk minimization in order to get good generalization speaking. Using l instances from two classes to identify the best hyper-plane necessitates using quadratic programming to solve an optimization problem.

$$(x_1, y_1), (x_2, y_2), \ldots (x_1, y_1), x_i \in R^N, x_i \in \{-1, +1\}$$

The margin between classes is the optimization criterion. A hyper-plane with discriminatory properties:

$$f(x) = \sum_{i=1}^{l} y_i a_i k(x, x_i) + b$$

A kernel function is defined as k(x,xi), with its value denoted by the sign of f(x). Making the best hyper-plane is the same as finding all AI that are not 0. A support vector of the ideal hyper-plane can be found at any data point xi that has an ai value greater than zero. Pedestrians and non-pedestrians are divided into two categories for classification purposes. Using the HOG descriptor, a feature vector is created, and this represents the feature points. Each of the feature points is given a weight value. The operating principle of SVM is to increase the distance between similar images. Because of this, SVM maximizes the separation between classes. Only a portion of the samples chosen as support vectors will be used in the hyper-development planes.

7.9 EXPERIMENTAL RESULTS

In this work the pedestrian images were classified by three ML approach. GLCM is used for feature extraction and SVM, NB, and RF for classification. Among the variety of existing ML algorithms, hence, the performance of SVM in pedestrian classification is validated by comparing its performance with NB based pedestrian classification. The results obtained by implementing traditional ML algorithms for pedestrian detection are discussed in subsections.

7.9.1 PERFORMANCE ANALYSIS OF SVM IN VISOR

This work validates the performance of ML classifier using the diverse performance measures like accuracy, precision, recall, and F-measure in VISOR dataset. Initially a contingency matrix consisting of TP, TN, FP, and FN values is constructed based on the performance of classifier.

7.9.2 EVALUATION MEASURE

The performance evaluation measures used to validate SVM based approach for pedestrian detection are precision, accuracy, recall, and F-Measure. These evaluation metrics are calculated from true positive (TP), true negative (TN), false positive (FP), and false negative (FN) values represented in the contingency matrix. The pedestrian detection system is trained to compute these evaluation metrics by comparing the actual data with the predicted data.

1. **Precision:** It is distinct as the percentage of designated images that are correctly categorized as pedestrians or non-pedestrians in class C out of all images in class C. It is intended using Eqn. (6) as given below:

$$\text{Precision} = \frac{tp}{tp + fp} \tag{6}$$

2. **Recall:** It's the proportion of accurate photographs in class C out of the total number of images in class C that are actually selected. It is calculated using Eqn. (7) as given below:

$$\text{Recall} = \frac{tp}{tp + fn} \tag{7}$$

3. **F-Measure:** It is distinct as the harmonic mean value of the proposed models precision and recall. It is calculated using Eqn. (8) as given below:

$$\text{F} - \text{measure} = \frac{tp}{tp + (fp + fn)} \tag{8}$$

4. **Accuracy:** Class C Prediction Accuracy Ratio is defined as the percentage of all properly classified predictions in class A, B, and C. It is determined using Eqn. (9).

$$AC = \frac{tp + tn}{tp + fp + tn + fn} \tag{9}$$

Table 7.2 represents that the performance analysis of three classifier under different parametric measures. In SVM classifier reached the better precision 93.45%, recall of 92.11% and F-measure of 91.36% respectively. RF achieved precision 92.82% and F-measure of 92.46%. NB achieved the precision of 92.10% and F-measure 93.54%, respectively.

TABLE 7.2 Performance Analysis of Different Classifier

Classifier	Precision	Recall	F-Measure
SVM	93.45	92.11	91.36
RF	92.82	92.75	92.46
NB	92.10	91.64	93.54

Figure 7.8 represents that the quantitative analysis of proposed methodology, in this comparison analysis, we used classifier with optimization and without optimization. In this performance comparison, PSO with classifier reached the better detection accuracy than without optimization technique.

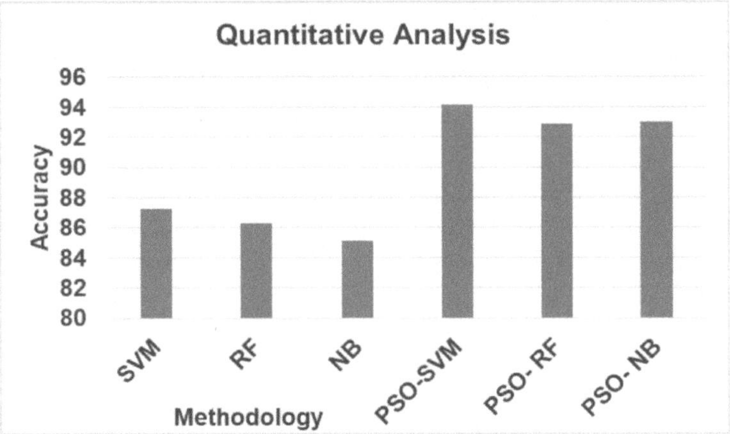

FIGURE 7.8 Quantitative analysis of proposed methodology.

Tables 7.3–7.5 represent detection accuracy of the three-classifier performance, in this analysis we takes 3,765 input images. In these total images, the SVM classifier reached the performance of images correctly detected as 3,544 and images in-correctly detected as 221. Another, the RF classifier reached the performance of 3,496 images are correctly detected and 264 images are detected incorrectly. Another classifier of Naïve Bayes classifier reached the performance of 3,501 images are correctly detected and 264 images are detected incorrectly.

TABLE 7.3 Performance Analysis of Detection Accuracy of SVM Classifier

Total Number of Input Images	SVM Classifier		
	Images Correctly Detected	Images In-Correctly Detected	Detection Accuracy (%)
3,765	3,544	221	94.13

TABLE 7.4 Performance Analysis of Detection Accuracy of Random Forest Classifier

Total Number of Input Images	Random Forest Classifier		
	Images Correctly Detected	Images In-Correctly Detected	Detection Accuracy (%)
3,765	3,496	269	92.85

TABLE 7.5 Performance Analysis of Detection Accuracy of Naïve Bayes Classifier

Total Number of Input Images	Naïve Bayes Classifier		
	Images Correctly Detected	Images In-Correctly Detected	Detection Accuracy (%)
3,765	3,501	264	92.98

Figure 7.9 represents that the graphical representation of three classifier performance under the proposed extraction and optimization concepts. In SVM classifier achieved the better accuracy of 94.13%, RF achieved the accuracy of 92.85% and Naïve Bayes classifier achieved the classification accuracy of 92.98%, respectively. In this comparison analysis SVM reached the better accuracy results.

7.10 CONCLUSION

In this present chapter, the role of ML approaches for the purpose of pedestrian detection is clearly discussed. In this proposed work the

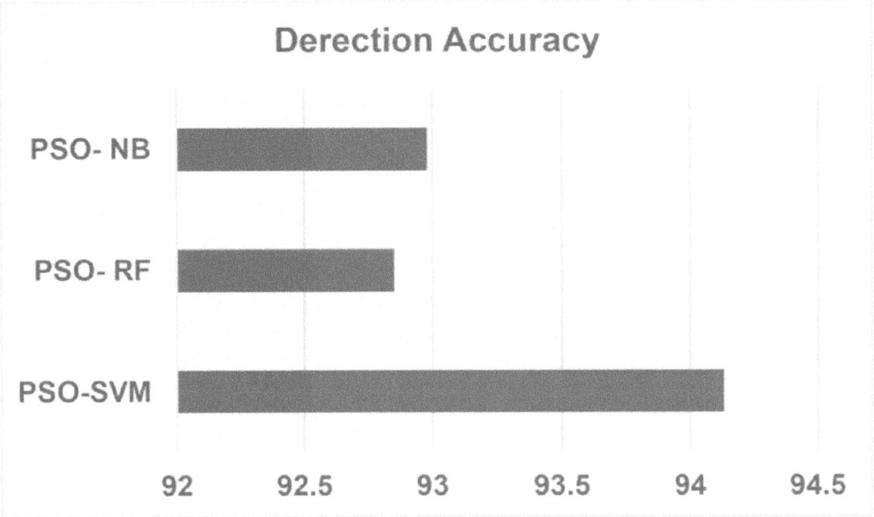

FIGURE 7.9 Graphical representation of accuracy comparison.

benchmark datasets from VISOR were used as input. Image frames are extracted image frames were analyzed for pedestrian classification. Analyzing the insights from the image data is entirely different from then other types of data as it involves number of features. From the experimentations done, the accuracy of ML models highly depend on the extraction of the healthy features from the set of available features. To enhance the accuracy in pedestrian classification it is necessary to concentrate on extracting the essential features from the entire feature set. In addition, the process of feature extraction has been addressed with the help of 8-Directional Gray Level Co-occurrence Feature Extraction based approach. In this proposed work initially, the pedestrian detection is done by using traditional ML algorithms like SVM, RF and Naïve Bayes (NB). To overcome the disadvantages of the traditional ML approach and hybrid ML approach, proposed research work concentrates on optimization approaches. PSO optimization is proposed to optimize the hyper-parameters of ML model. The results and performance evaluation shows that the optimized ML classifier to produce the maximum accuracy in detecting the pedestrians are compared with three algorithms, respectively.

KEYWORDS

- **autonomous vehicles**
- **data augmentation**
- **hyper parameter**
- **machine learning**
- **optimization**
- **pedestrian detection**
- **support vector machine**

REFERENCES

1. Gavrila, D. M., (2000). Pedestrian detection from a moving vehicle. In: *European Conference on Computer Vision* (pp. 37–49). Springer, Berlin, Heidelberg.
2. Zabłocki, M., Gościewska, K., Frejlichowski, D., & Hofman, R., (2014). Intelligent video surveillance systems for public spaces–a survey. *Journal of Theoretical and Applied Computer Science, 8*(4), 13–27.
3. Zhang, J., & Tao, D., (2020). Empowering things with intelligence: A survey of the progress, challenges, and opportunities in artificial intelligence of things. *IEEE Internet of Things Journal, 8*(10), 7789–7817.
4. Fang, Z., Jiang, Y., Xu, H., Shaw, S. L., Li, L., & Geng, X., (2018). An invisible salient landmark approach to locating pedestrians for predesigned business card route of pedestrian navigation. *Sensors, 18*(9), 3164.
5. Ivanchenko, V., Coughlan, J., & Shen, H., (2008). Crosswatch: A camera phone system for orienting visually impaired pedestrians at traffic intersections. In: *International Conference on Computers for Handicapped Persons* (pp. 1122–1128). Springer, Berlin, Heidelberg.
6. Dai, C., Zheng, Y., & Li, X., (2007). Pedestrian detection and tracking in infrared imagery using shape and appearance. *Computer Vision and Image Understanding, 106*(2, 3), 288–299.
7. Wang, D. Z., Posner, I., & Newman, P., (2012). What could move? Finding cars, pedestrians, and bicyclists in 3D laser data. In: *2012 IEEE International Conference on Robotics and Automation* (pp. 4038–4044). IEEE.
8. Wu, H., Zhang, X., Story, B., & Rajan, D., (2019). Accurate vehicle detection using multi-camera data fusion and machine learning. In: *ICASSP 2019–2019 IEEE International Conference on Acoustics, Speech, and Signal Processing (ICASSP)* (pp. 3767–3771). IEEE.

9. Xie, S., Zhang, X., & Cai, J., (2019). Video crowd detection and abnormal behavior model detection based on machine learning method. *Neural Computing and Applications, 31*(1), 175–184.
10. Zheng, G., & Chen, Y., (2012). A review on vision-based pedestrian detection. In: *2012 IEEE Global High Tech Congress on Electronics* (pp. 49–54). IEEE.
11. Rozsa, Z., & Sziranyi, T., (2018). Obstacle prediction for automated guided vehicles based on point clouds measured by a tilted LIDAR sensor. *IEEE Transactions on Intelligent Transportation Systems, 19*(8), 2708–2720.
12. Zhang, C., Rui, Y., Crawford, J., & He, L. W., (2008). An automated end-to-end lecture capture and broadcasting system. *ACM Transactions on Multimedia Computing, Communications, and Applications (TOMM), 4*(1), 1–23.
13. Zigel, Y., Litvak, D., & Gannot, I., (2009). A method for automatic fall detection of elderly people using floor vibrations and sound—Proof of concept on human mimicking doll falls. *IEEE Transactions on Biomedical Engineering, 56*(12), 2858–2867.
14. Gandhi, T., & Trivedi, M. M., (2007). Pedestrian protection systems: Issues, survey, and challenges. *IEEE Transactions on Intelligent Transportation Systems, 8*(3), 413–430.
15. Li, H., Wu, Z., & Zhang, J., (2016). Pedestrian detection based on deep learning model. In: *2016 9th International Congress on Image and Signal Processing, Biomedical Engineering, and Informatics (CISP-BMEI)* (pp. 796–800). IEEE.
16. Wang, Y., Piérard, S., Su, S. Z., & Jodoin, P. M., (2017). Improving pedestrian detection using motion-guided filtering. *Pattern Recognition Letters, 96*, 106–112.
17. Kim, J. H., Batchuluun, G., & Park, K. R., (2018). Pedestrian detection based on faster R-CNN in nighttime by fusing deep convolutional features of successive images. *Expert Systems with Applications, 114*, 15–33.
18. Llorca, D. F., Sotelo, M. A., Parra, I., Ocaña, M., & Bergasa, L. M., (2010). Error analysis in a stereo vision-based pedestrian detection sensor for collision avoidance applications. *Sensors, 10*(4), 3741–3758.
19. Dollar, P., Wojek, C., Schiele, B., & Perona, P., (2011). Pedestrian detection: An evaluation of the state of the art. *IEEE Transactions on Pattern Analysis and Machine Intelligence, 34*(4), 743–761.
20. Hernandez-Belmonte, U. H., & Ayala-Ramirez, V., (2016). Feature selection using genetic algorithms for hand posture recognition. In: *Mexican Conference on Pattern Recognition* (pp. 208–218). Springer, Cham.
21. Zhang, C., Li, H., Wang, X., & Yang, X., (2015). Cross-scene crowd counting via deep convolutional neural networks. In: *Proceedings of the IEEE Conference on Computer Vision and Pattern Recognition* (pp. 833–841).
22. Walach, E., & Wolf, L., (2016). Learning to count with CNN boosting. In: *European Conference on Computer Vision* (pp. 660–676). Springer, Cham.
23. Xiao, T., Li, S., Wang, B., Lin, L., & Wang, X., (2017). Joint detection and identification feature learning for person search. In: *Proceedings of the IEEE Conference on Computer Vision and Pattern Recognition* (pp. 3415–3424).
24. Jaderberg, M., Simonyan, K., Vedaldi, A., & Zisserman, A., (2016). Reading text in the wild with convolutional neural networks. *International Journal of Computer Vision, 116*(1), 1–20.

25. Tomè, D., Monti, F., Baroffio, L., Bondi, L., Tagliasacchi, M., & Tubaro, S., (2016). Deep convolutional neural networks for pedestrian detection. *Signal Processing: Image Communication, 47*, 482–489.
26. Bella, M. I., & Vasuki, A., (2019). An efficient image retrieval framework using fused information feature. *Computers & Electrical Engineering, 75*, 46–60.
27. Guo, X. C., Yang, J. H., Wu, C. G., Wang, C. Y., & Liang, Y. C., (2008). A novel LS-SVMs hyper-parameter selection based on particle swarm optimization. *Neurocomputing, 71*(16–18), 3211–3215.
28. Premebida, C., Ludwig, O., & Nunes, U., (2009). Exploiting LIDAR-based features on pedestrian detection in urban scenarios. In: *2009 12th International IEEE Conference on Intelligent Transportation Systems* (pp. 1–6). IEEE.

CHAPTER 8

Smart Railway Cargo Management System Using Big Data

ADITI LAXMI GIRI, A. JOHN, and E. RAJESH

School of Computing Science and Engineering, Galgotias University, Greater Noida, Uttar Pradesh, India

ABSTRACT

In this chapter, we will discuss and try to understand Big Data followed by is implementation in smart railway cargo management system. We will look into things with a practical aspect by considering real world examples. Before understanding the use of big data in railway cargo management system, we will focus on understanding what big data is in the first place and why it is trending nowadays. This includes addressing its characteristics and its types which is essential to keep in mind while working on it. In order to be familiar with the progress in the field of big data, one has to dig deep into its history. As data in involved in every aspect of our life, similarly big data is involved in almost every domain of Industry 4.0. It has a huge connection to it and with time it has become one of the most basic and fundamental part of it which acts as the core or foundation for many other technologies.

As far as railway cargo management system is concerned, the foremost thing we will try understanding is the impact of Industry 4.0 on the transportation system. The developments done and required enhancements for coping up with the present day commercial and industrial markets. Acknowledging the objectives of railway cargo management system using big data and the need for its adoption is the key to understanding

Smart Mobility and Intelligent Transportation Systems for Commercial and Hazardous Vehicles.
T. Poongodi, Naga Swetha Pasupuleti, and Balamurugan Balusamy (Eds.)

its architecture and working in a better and easier way. As per the current status of rail line cargo information resources, the general design of rail route cargo huge information application dependent on the rail route information administration stage is proposed, including the handling streams of information obtaining also, access, resources catalog development, information displaying and the board, information stockpiling and investigation, and so forth.

By recognizing the benefits big data provides to railway cargo management system, we can work towards its implementation in a better manner. But at the same time, it is important to address the challenges and issues as well which might affect the quality of resultant data. Moreover, having a brief idea of its future scope helps in developing one's interest in a particular subject to a greater extent.

8.1 INTRODUCTION

8.1.1 WHAT IS BIG DATA?

Evolution in the field of computer science or information technology (IT), and its usage in most of the other fields such as medical science, education, business, agriculture, etc., has made data to grow at an exceptionally tremendous rate. With the ongoing Fourth Industrial Revolution, often known as "Industry 4.0," which focuses on the automation of the traditional manufacturing, industrial practices and automated management system, the traditional ways of managing and processing data do not fit into the current scenarios remarkably.

As IT is growing rapidly, tons of data are being generated digitally as well as on the Internet on a regular basis. The problem of managing a large scale of data did not occur suddenly. It has been going on for quite a decade now. However, in general, over the years it has been a known fact that the creation of data is much easier than managing it or analyzing it in order to get some meaningful results. Analysis of data plays a crucial role in most of the new age trending technologies such as machine learning (ML), internet of things (IoT), robotics, and data science models. Although the computer systems which are available today are much faster than they were in the 1930s, the enormous amount of data that we have today brings in a lot of pressure in the analysis procedure using these computers as well.

As a result, in order to address these problems related to large-scale data analysis, many effective methods were demonstrated some of which are modeling, data summary, density-based approaches, grid-based approaches, divide, and conquer and distributed computing. All these had one thing in thing, that is, their main aim was to improvise the performance of the operators of the data analytics process. After a lot of analysis and research, the finals results suggested that with proper and effective methods, we can analyze large-scale data without much hassle or inconvenience, in a reasonable period.

A specific example is the dimensionality reduction method, which aims to reduce the amount of input data to speed up the data analysis process. Another reduction method model that reduces the data calculation of data clustering, which can also be used to speed up the calculation time of data analytics. Big Data was a term that was given to huge amount of data and the problems related to large-scale data such as its management and analysis. In general, big data refers to the data that is generated on a large scale and is usually dynamic in nature, i.e., increases at a considerably higher rate over the time. Its growing nature is one of the major reasons why the previously existing data management tools do not work efficiently with this new form of data.

Big data is often identified by 3V's which are velocity, volume, and variety, which often termed as its characteristics. The data in the data warehouse is collected from various resources for analysis and discovering patterns in order to derive useful information resulting in knowledge. As a result of this, data from multiple resources collectively becomes varied. This is referred to the variety of data. Volume is considered because large data will give more accurate results. The rate of growth of data with time is known as its velocity. Apart from the major three characteristics described above, there are other aspects as well which can be considered for a better definition of big data. These are veracity, value, variability, and exhaustiveness of the data. Veracity focuses on the reliability and accuracy of data that is obtained. It is a necessary aspect to consider while data cleansing, matching, drawing patterns, etc. Value of the data means its importance. Variability as the name suggests refers to the different format or structure of data.

People often get confused with two of the most commonly used terms: Big Data and Business Intelligence. Business Intelligence focuses on implementing analytical tools and is concerned with statistics in general.

Big Data uses these along with some optimization and relationship prediction algorithms to derive some meaningful information from a large pool of data. It also focuses on predicting possible outcomes by looking at the available dataset as an input. Therefore, it is considered by many industries and businesses. Big data is used in all the disciplines of computer science which are concerned with large datasets [1, 2].

8.1.2 HISTORY OF BIG DATA

Although the concept of big data might seem to be very new, its idea first emerged in the 1960s and 1970s. Data was first introduced with the development of data centers and relational databases during this period. Although it is not accurately known who gave this term in the first place acknowledge John R. Mashey for it. Data analysis and analytics is not a very new thing. People have been trying to use them for support and betterment of the decision-making activities for centuries now. Around 300 BC, the ancient Egyptians attempted to encapsulate the entire data available in Alexandria's library. Furthermore, Roman Empires analyzed statistics related to their military for ideally distributing their armies in different regions. Data of one form or the other is used in almost every aspect of our everyday life.

However, with the digitalization and computerization of almost everything over the past few decades, the amount of data that is created has gone far beyond human embodiment. This problem became more critical in 2005, when software professionals started to see the volume of data being produced regularly by Facebook, YouTube, and other commonly used online services. This brought with itself the need for a new system or method which was able to fit into the current scenarios and can easily cope up with the increasing rate of data. To clearly understand the development and evolution of Big Data, we can broadly divide it into three phases where each phase has its significance in terms of characteristics and abilities. In an attempt to understand and explain the meaning of Big Data in today's world, it is very crucial to know the importance of each of these phases and how it contributes to the contemporary definition of big data [3].

- **Big Data Phase 1.0:** Data analysis or data analytics and Big Data were originated from a broader domain commonly known as database management. It majorly focused on the retention, retrieval, and optimization techniques which is common for data in relational database management systems (RDBMS). Data management and warehousing were the core of this phase. Big Data Phase 2.0: This is linked with the rise of the Internet in the early 2000s, which expanded the opportunity for data collections and analysis. Expansion of e-commerce websites gradually increased the need of analyzing customer behaviors. HTTP web traffic established a huge rise in semi-structured as well as unstructured data.

- **Big Data Phase 3.0:** Apart from the content being generated from the web, mobile devices also started playing a crucial role in producing valuable information. With the emergence of mobile devices, the possibility of analyzing behavioral data, tracking movement and even health-related information grew to a wider extent. This kind of data provides a wide range of opportunities for innovations in different fields. The development of open-source big data frameworks such as Hadoop (and recently, Spark) is essential for the growth of large data because they make it cheaper to operate and store large data. In the following years, the size of big data has skyrocketed. Users are still generating large amounts of data – but now it's not just humans. In the era of robotics, artificial intelligence (AI), and ML, machines themselves are capable of developing results from their own produced datasets. With the advent of the IoT, more and more devices and devices are being connected to the Internet to collect data on customer usage patterns and product performance. Big data has quite a long way to go, and its current usage is just the beginning of it. Cloud computing is another fundamental technology which further expands the possibilities of big data. Cloud technology provides elastic scalability, where developers can rotate temporary clusters to test subsets of data. Big data and cloud computing are a perfect combination and go hand in hand. Cloud computing deals with storing and managing a huge amount of data stored remotely which in turn is connected with the principal idea of big data. Repositories of big data can exist in many forms depending upon the needs and desired outputs [4].

8.1.3 TYPES OF BIG DATA

With the surge of the Internet age, each second an unfathomable amount of data is being created. Data can be anything like a comment on a post, a list of liked videos on YouTube, search history, pictures, etc. Hence, it is very clear that all the data that is created is not of the same type. This is also one of the main reasons why analysis and cleaning of data are considered to be a strenuous task.

The first step for all the data management and data mining process is the selection of data as per requirement which is followed by a grouping of similar type of data together. Grouping of similar type of data is often done on the basis of their structure, complexity, and the technique or tools required for examining it. This is done because the same method or tools cannot be applied to all kinds of documents. The hierarchy structure of the classification of big data shown in Figure 8.1. Hence, big data is broadly classified into four types which are:

- Structured;
- Semi-structured;
- Unstructured; and
- Quasi-structured.

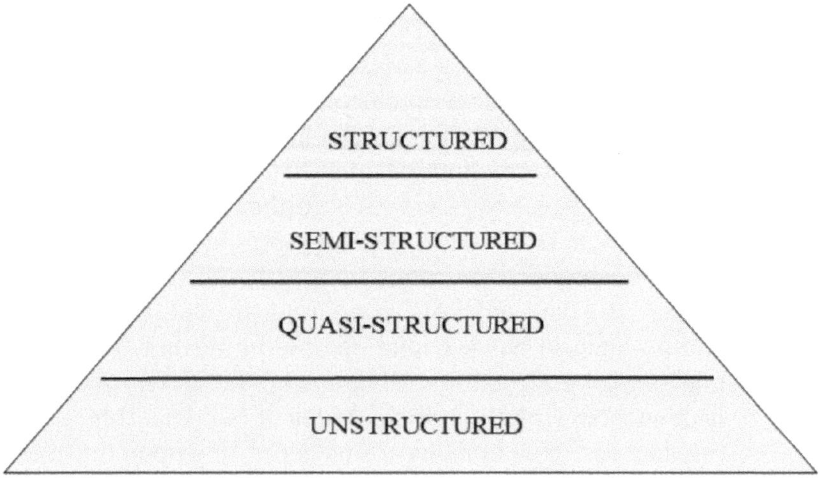

FIGURE 8.1 Hierarchy structure of the classification of big data.

The philosophy of big data encompasses these four types of data, but the main focus is laid on unstructured data. These three terms are applied technically at all levels of analysis in big data. Understanding the source of raw data and how to treat it before moving on to the analysis part is very important. Hence, the data extraction process must be effective enough to forge the effort worthwhile.

1. **Structured Data:** It refers to the type of data which can easily be stored, retrieved or processed in a particular fixed format. It is usually stored in an ordered way which is why it is easiest to work with. Structured knowledge features a dedicated knowledge model, that conjointly features a well-defined structure following a coherent sequence and is intended such it's simply accessible and usable by a personal or laptop. Structured knowledge is sometimes held on in well-defined columns and databases. Data stored in database management systems (DBMS) can be considered as its example.
2. **Semi-Structured Data:** It refers to data that does not have a proper schema which is why it has a structured form that is not well defined. Therefore, it is said to be a combination of structured as well as unstructured data. It inherits some options of structured knowledge; however, a serious part of such knowledge fails as a static structure and concurrently, it doesn't follow the formal structure of information models like RDBMS. Example of such data would be CSV, TSV, online transaction processing, etc.
3. **Unstructured Data:** It is one which includes all kinds of audio, video, log, and images files as they do not have a fixed model or structure. Working with this type of data is regarded as the most difficult task, especially when it is in large volumes. It is flexible though because of the lack of schema.
4. **Quasi-Structured Data:** It consists of textual data which has inconsistent data formats that can be formatted using various tools. Web server log which is a log file that is created by some server is an adequate example of quasi-structured data.

8.1.4 INDUSTRY 4.0

The current Fourth Industrial Revolution, often referred to as Industry 4.0, aims to automate the traditional industrial and manufacturing practices with the help of smart technologies present in the modern world. This is generally done by integrating machine-to-machine communication (M2M), embedded systems and the IoT in order to achieve increased automation levels.

This phrase was first given by a group of scientists, working for the German government, to develop a high-tech strategy. Afterwards, it was introduced to a wider audience by Klaus Schwab, who was then the executive chairman of the World Economic Forum. He published the same in an article by Foreign Affairs in the year 2005. Later on, in 2016, "Mastering the Fourth Industrial Revolution" was made as to the theme for the World Economic Forum Annual Meeting in Switzerland. The First Industrial Revolution was characterized by the adaptation of machines above the manual production method with the help of power generated through steam and water. This took a long period of time which mainly affected the textile manufacturing process, metal industries, mining process and more. The Second Industrial Revolution is otherwise known as Technological Revolution as it focused on implementing and establishing networks using electricity. This was followed by the Digital Revolution or Third Industrial Revolution which began in the late 20th century after the end of two world wars. Digitalization was much needed at that time due to the ultimate drop-off in industrialization and technical processes [5, 6].

So, basically, we are in the middle of a significant change which is so important that it has been termed as Industry 4.0. Ages after the mechanization of the commercial world by water and steam power to mass manufacturing and laying assembly lines for establishing networks followed by digitalization, we have already come quite a long way. The next major step according to Industry 4.0 is to minimize human interference and ease the processes, by taking forward the trend of the Third Industrial Revolution and complimenting it with smart and automated systems which are powered by data science and ML. As a result, it optimizes the digitalization process start in Industry 3.0. The designing of Industry 4.0 is mainly based on four principles which include establishing interconnection, maintaining lucidity of information, providing technical assistance in burdensome tasks and decentralization of decision for cyber-physical systems. Some of the biggest trends within this are robotics, additive manufacturing or 3D printing, smart homes and appliances using the IoT, cloud computing, development of sensors and development of different prediction systems [7].

8.1.5 USE OF BIG DATA IN INDUSTRY 4.0

The fourth Industrial Revolution refers to the integration of the real world along with the virtual world. The digital revolution has been marked by technologies that take advantage of big data and AI to nurture automatic learning systems. Makers in today's marketplace ask to attain business intelligence through the compilation, analysis, and sharing of information across all key practical domains so as to obtain excellence in production.

Businesses ought to method information into timely and valuable data for decision making/deciding/higher cognitive method and process optimization activities. Today's competitive business atmosphere forces enterprises to method high-speed information and integrates valuable data in production processes. For instance, the conception of Industry 4.0 is predicted to vary production within the close to future. During this conception, machines during a good industrial plant move with their environments. Standard machines remodel into context-aware, conscious, and self-learner devices. This transformation provides these devices with the aptitude to method period of time information to self-diagnose and forestall potential disruptions in the production method. What is more, once such a machine is appointed a task, it will self-calibrate and grade between tasks to optimize production quality or potency. The volume and velocity of data being generated will only grow more and more in the upcoming years due to the widespread use of IoT devices and sensors. With almost all the sectors adopting technologies like edge computing, cloud computing, fog computing, etc. Industry 4.0 has successfully made its way to the ubiquitous use of big data analytics. Industries are looking for ways to collect, store, create, and process data from various sources [8].

The following are some of the uses (or need) of big data in the current evolution:

1. It reduces the chances of breakdowns and unpredicted downtime as big data analytics can be used to discover patterns thereby

predicting the possible outcomes. This helps in decreasing the chances of failures.

2. It helps in integrating the quality data with production data whose resultant (produced after proper analysis) can be used to enhance the quality of production.
3. Big data recognize the factors which affect the performance, that too with no additional cost. It helps in getting an idea of all the loopholes within the system and therefore eliminates bottleneck.
4. Helps in brushing up or easing the warehouse activities by detection of errors, quality control analysis and suggesting optimal solutions.
5. Data that is produced every minute by IoT devices or sensors are collectively so huge that they cannot be managed without the use of big data analytics.
6. In order to implement machine learning and deep learning (DL) algorithms, it is essential to take into account a large amount of data as input as accuracy is directly proportional to the volume and quality of data fed as the input.
7. Big data analytics is more or less identical to predictive techniques as it gives a brief idea of what might be the possible output.
8. With systems being automated, devices and machines are becoming capable of creating data by themselves each second. This data cannot be ignored or omitted. It has to be stored properly for not only finding patterns but also for maintaining records.
9. Innovation and research of self-driven or automated cars, a huge volume of data will be required for maintaining the records of each individual car, proposing shortest routes, predicting maintenance, and more.
10. Big data is a major part of cloud computing because the large scale is to be stored, maintained, and make easily accessible. These are also prone to constant changes as per client demands.
11. Big data or big data analytics acts as the core of automated factories or automated production management in general.
12. It is also an essential part of developing effective business strategies.

Technologies of the contemporary world would have never been able to decrypt hidden information and accurate values from large data pools without the use of big data. It is the core for predicting patterns and extracting meaningful information from raw facts which is further known as knowledge. Hence, Industry 4.0 would not have existed nor any of the smart technologies in the absence of big data [9].

8.1.6 AUTOMATION OF MANAGEMENT SYSTEMS

Without automated management systems, companies would not have been able to cope up with growing pressures of availability and consistent continuity. It is the only way to ensure that their systems are working at an optimized rate to meet the growing demands. Data is one of the most important virtues for any organization. With the help of automated management systems, organizations can reduce the complex management tasks by 1.5 times and can also free up the resources at the same rate.

Many small-scale businesses or startups fear delegating the tasks to automated systems more than doing so to their employees. When in reality, they are the ones who can have more benefits from such systems. Automated management systems might seem to be expensive at first but on proper calculation, it can be seen and understood that they are more efficient than the traditional management techniques. Reason being that it reduces the need for human resources, improves accuracy and provides results in less time.

Nowadays, as computer systems are spreading all around the world and are involved in every field due to digitalization, the networking system has become more complex which lays constant pressure on achieving higher performance rates, better services and increasing reliability. The basic problem of managing on a large scale has been intensified due to these demands. As a result, it is very common that few of the regular or repetitive, yet important tasks might be overlooked when managed by humans. Therefore, in order to provide the most efficient results to the clients, organizations require a solution that can easily be deployed without consuming much time and causing chaos [10].

The basic idea behind creating effective management systems is that they should be able to predict the failures and provide a warning prior to the occurrence of such events. It should also be automated enough to handle the basic tasks with the need for any human intervention. This is not something that can happen instantaneously. Moreover, even though the basic idea or requirements are the same, every system requires customized support. The same software has to be modified in different ways for different businesses and organizations as per their demands. Using automation, we can easily monitor and manage disk spaces, system notifications, events, performance, and job queues in an effective and cost-efficient manner [11].

Following are some of the benefits of automating management systems:

- Organizations can spend more time on strategic execution than on administrative tasks.
- It reduces expenses by reducing required human resources.
- Manages the performance of workflows by customizing processes and hence making them seamless.
- Provides instant feedback and error notifications.
- Decreases the dependency on physical documents.
- Identifies vacant resources by betterment management of disk spaces.
- It can easily be managed even when decentralized.
- Helps in developing new and advanced business opportunities.
- Provides security and restricted accessibility.

However, everything comes with its own set of challenges. One of the most important and crucial among them is the deployment of these systems and their integration with the organizations existing systems so that it causes minimum inconvenience during the deployment period as well as afterwards. Sometimes this included altering some of the basic structures and features of the developed automated system as per the organizations' needs. Moreover, not all automated systems are less costly than the amount spend when the same work is done by humans. Hence, it important to analyze the needs and budget before getting any automated system. Complexity and security threats due to hacking are also some of the major challenges. Some of these systems are so complex that they might require trained individuals for operating them. However, a well-automated system is one with less complexity as that is its main job, to reduce the complexity, but it entirely depends on the needs and type of business or organization where it is being used [12].

8.2 SMART RAILWAY CARGO MANAGEMENT SYSTEM USING BIG DATA

It is clear from the above discussion that the current industrial revolution has quite a different approach based on smart technologies for the existing management format. In order to obtain a leading position in today's market, it is essential to utilize the opportunities of digitalization to the greatest. This change has been quite recognizable in the transport business. Telematics systems, which include position detection abilities using GPS, have been used in road cargo transportation for quite a while now. Data mining or big data analytics has been playing a crucial role in preparing the air traffic architecture, developing optimal strategies, processing related data and laying networks. IoT has already become an important part of many industries including freight.

Railway industries all around the world have shown a keen interest in big data analytics. It is mainly because of the wide use of railways as a means of transportation due to which a gigantic amount of data is produced. This data not only includes the details of the passengers but also the details of all the important work which does not meet the eyes at first sight. It contains details about the staff, from the higher officials to the ones doing the most basic jobs such as cleaning or catering. Each and every detail is to be kept with utmost care for security reasons. As the Indian Railway System is the largest employer in the country, imagine the amount of data that is produced by it every second. What if all this data had to be managed traditionally? It is just impossible to manage such enormous data without the help of any computerized data management system. Not just in India but all around the world, people mostly use railways for traveling when the distance to be cover is of few hours by rail. Hence, railway systems worldwide had to opt for the digitalization trend growing within this industrial revolution [13].

This was not the end of this problem. Such huge volumes of data required powerful software which was capable of storing it as well as provided ease in accessibility and modification or working on it. This was not possible with the normal database management systems and therefore a similar, yet advanced approach was needed. This is when big data came into the picture. The very first idea of considering big data in railway management system was introduced by Faulker [18]. It divided the data produced in railway system into broadly four types:

1. The basic data for configuration is static in nature and reflects real-world entities. It changes only when some maintenance work is done or some modification in any of the entities as a part of some serious updates.
2. Complexities of scheduling trains define the type of infrastructure that is required.
3. Status data includes a wide range like the status of reservation, status of train, status of employees doing various jobs, status of work, etc. This was often required by the external or mobile reporting systems using a different interface.
4. Data regarding operation carried out individually which is usually given to the railway control system manually.

The increasing demand brings with itself the need of providing data confidentiality and accessibility for higher customer satisfaction. Big data analytics now holds a large part in the decision-making procedure of railway systems. It can distinguish the conduct of bottlenecks, most extreme burdens, variety in traffic, impromptu postpone timings, Big Data in Railways examination timings and mishaps that will affect client's solace, business' misfortunes and resources notoriety. Despite all its advantages, many of the railway companies have not implemented it yet. Few reasons behind this are the lack of knowledge on how to integrate it within the system, difficulty in obtaining and analyzing such large-scale data, security issues and routinization. However, no matter how hard it may sound, it is actually very easy to assimilate and use when done with proper planning and with professional help [14].

Considering rail cargo, as an industry, regarding flexibility to the necessities of the consumer, actually stay behind rivals in this monetary circle. The client expects that data about shipping cargo will be consistently close by it implies that data will be accessible on the site as well as in the versatile application. Few out of every odd organization in the European region ready to give such a volume of data as rivals on the lookout. The driver of the train has no data about the load during transportation, so the entirety of the issues that have emerged through the course will be distinguished uniquely at the focuses where the cargo will be dumped from the carts. The genuine condition of refreshing data about transportation additionally is low, it brought about by the way that the data is gone into the product just upon the appearance of the train at the freight depot.

8.2.1 *OBJECTIVES*

The goal of big data in railroads is to empower prescient calculations from heterogeneous information sources, adaptable information structures, constant correspondences, and perception techniques. Such exploration is trying to framework resource upkeep in the rail route climate in three specific zones: railroad framework segment debasement forecast displaying; rail line foundation and vehicle support cost expectation demonstrating, and foundation and checking vehicle's current state. The goals of big data investigation in the railroad business are as per the following:

- Create continuous prescient calculations from heterogeneous information sources that will adapt to protection safeguarded preparing, highlight, and occurrence determination, discretization, information pressure, outfit classifiers and relapse models, and spatial and transient arrangement of information.
- Create adaptable information structures dependent on cross-space information source procurement through a virtualization layer between the information securing interaction and information inspection. This will incorporate new arrangements that consolidate new information base capacities to coordinate heterogeneous information sources in an elite getting to framework dependent on the cloud service.
- Enable large information interchanges utilizing open interface passages with checking frameworks giving timestamp and position synchronization, heterogeneous correspondence support, including portability and conglomeration, and need conventions for ongoing transmission of data.
- Apply representation strategies of infographics and virtual/ increased reality.

8.2.2 *NEED FOR ADAPTATION*

Surveys have shown that there is a mutual discernment among the frequent rail users that the railway's sector does not desire to evolve. The reason behind this impression is the lack of flexibility in its organization that is knowledgeable about participation. Generally, a normal potential client has insight into the method of working of the road system and isn't mindful of the intricacy hidden in the rail framework. Then again additionally, the impression of the rail area of the potential rail client is frequently not positive. In numerous events, it is referenced that the primary justification for not utilizing rail is that one is 'simply not all around educated about the prospects of rail' or 'they would prefer not to see how it functions.' One of the basic reasons for this discernment is that while surveying the rail execution a coherent and for the most part acknowledged methodology is to take a gander at the current rail cargo administrations and their present customers. As a rule, the consequences of such an appraisal are that the customers are happy with the help given. Notwithstanding, for the rail area a significant goal is to pull in another objective gathering which is right now utilizing street transportation (making a shift to rail). This objective gathering has various prerequisites which the rail area should offer to make it feasible for them to utilize rail [15].

Generally, railroads cargo are viewed as a method for moving huge volumes of mass load over significant distances. Their capacities could accordingly be viewed as independent with next to no consideration regarding the significance of different pieces of the production network. In any case, the attention on the new freight to be pulled in which is at present moved out and about, in the mix with the overall advancement of worldwide stockpile chains, powers the rail route area' to adjust its perspective. Concentrated inventory network direction of the rail line area is required for improving the administrations and serious situation of the rail line area. While supply chain management (SCM) centers around the administration of trade streams inside and across the individuals from the inventory network, supply chain orientation (SCO) underscores the essential mindfulness and embracing of SCM inside an individual inventory network entertainer. All the more exactly, SCO says that the acknowledgement by an association of efficient and vital ramifications of the strategic exercises are engaged with dealings in an inventory network. This implies that the SCO idea accepts that the powerful execution of SCM needs an entertainer to put key accentuation on the reconciliation important to work with SCM trade. This idea envelops the idea of inventory network combination, which alludes to the degree to which separate gatherings cooperate agreeably to show up at commonly satisfactory results. Store network direction hence utilizes builds and estimations relating to the level of participation, coordination, connection, and joint effort. Digitalization in cargo system is the key to achieving the smart supply chain.

8.2.3 REQUIRED DATA

In order to maintain a smooth SCO, the railway cargo department needs to look into SCM techniques and their betterment, which is marked as a fusion of business processes such as handling client relationship, service administration, demand fulfillment and marketability. The higher the extent of freight or cargo in SCO, the higher will be the extent of integration of all the process during deployment over its network. Data is an important aspect of the interaction process among the collaborators. Therefore, to achieve the best services while managing demand and supply chain and have proper control over them, some basic information should be kept in mind few of which may include:

- The maximum capacity during each trip of each train/rail;
- The available of railway routes;
- Whereabouts of the locomotives and wagons, and distance between them;
- Arranging the wagons for loading and unloading with ease while ensuring the prioritization constraint of the goods;
- Minimization of unscheduled waiting time;
- Timely maintenance;
- Proper communication among the employees;
- Departure time, also with regards to the real-time scenario;
- Arrival time, also with regards to the real-time scenario.

To move to an inventory network direction of the rail area and make benefits that meet the measures of the current street transportation clients, the progressions should come from the rail area and not its customers. In any case, a superior comprehension of store network necessities ought to be made. Further, a more express division of the rail cargo market will likewise improve the correspondence on the required improvement of this new market fragment and the distinctions with the customary rail cargo market sector.

In the progress of transition, it is imperative to survey the rail route area's inventory network directly from the point of view of the rail route clients and different entertainers in the store network. Assessing the degree to which transport clients and other non-rail line related partners see rail routes to be inventory network arranged could prompt determinations whether there is a union in their discernments. An excess hole in insights may demonstrate either the chance of over accentuation in fragments that clients don't esteem or an expected shortage in the arrangement of administrations in portions they do.

8.2.4 ARCHITECTURE

Before moving on to the architecture, we need to understand how the data for cargo asset is managed. Analysis of structured data consists of analyzing multi-tables, domains, views, etc., in the database. Contrarily, in case of unstructured data analysis of the storage type, many things like data format and description, finding frequency, etc., are involved. Asset management incorporates describing and defining the status of cargo data on the basis of name of the system, department that is in charge of the particular system, its deployment level, data inventory, data incrementation either monthly or in any selected time. During all this, the most important thing is to keep in mind the concern department whose asset is being considered. It can be either regular railway management system data or railway cargo management system data [16].

The utilizations of cargo management system using big data take information assortment and administration as the advancement point, with data inspection and mining as the center to grow big data applications identified in cargo transportation. These applications help in successfully integrating all the related information like the cargo station details, information regarding real-time tracking, equipment required, data related to logistics and even more. Furthermore, these applications are the ones responsible for fabricating cargo information models for other uses such as business examination, updations, and improvement of the center elements of the rail route information administration stage, constructing a particular cargo demonstrating calculation library to frame information application investigation results to show on or push to existing cargo business frameworks, offering specialized help for resource worth and the board range of options. Consider the following diagram for better understanding big data application in cargo.

1. **Data Collection and Accessibility:** Railway cargo data sources consist of data from some particular business system and data generated outside the organization like by the customers. Business framework information alludes to information from cargo claims to fame, for example, cargo vehicle global positioning frameworks, cargo vehicle load clearing frameworks, and confirmation frameworks, and so forth. Outside data alludes to information identified with rail route cargo transportation like local financial information, meteorological information just as extensive traffic information, and even more. As per the prerequisites of cargo information assortment recurrence, information with high ongoing necessities or the one with higher priority should be gathered progressively by using Message Queues, Web Services, RESTful APIs, etc., while data other than the real-time data is gathered and clumped together. It can be consistently transferred and communicated by different means such as JBDC/OBDC, FTP/SFTP, and more. This can be done to meet the entrance necessities of heterogeneous information sources like social data sets and non-social data sets.
2. **Data Resource Inventory:** The data which is gathered has to be investigated with information arrangement to explain the dissemination of information assets in a better and optimal manner. It simply means that to set up a planning connection between the information tables of the business frameworks and the information models, and structure a progressive order inventory, that is hierarchical categorization, of the information in order to give a bound together resource view to information application, endorsement, and sharing, accomplishing an all-encompassing showcase of cargo information to guarantee the successful uprightness and information resource utilization.
3. **Data Modeling and Administration:** In this step, subject-oriented strategy is embraced for constructing the data model for railways cargo management system, coordinating the data collected from various business processes, utilizing a bound together legitimate language as a means for depicting entities and associations among them by maintaining the data integrity and consistency constraints. Data cleaning is an essential step that is done to remove noisy data by elimination of errors and

redundant data. Next is the administration process which incorporates standard administration and quality administration. Standard administration focuses on giving general operational options like browsing and acquiring whereas quality administration alludes to Figure 8.2 assigned information observing and evaluating rules to consistently improve information quality.

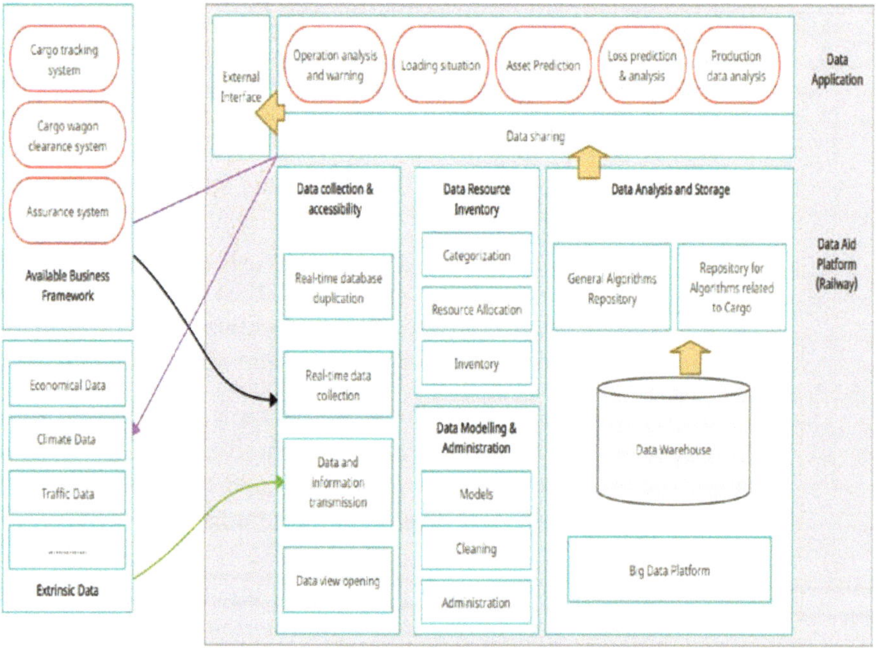

FIGURE 8.2 Architecture of cargo tracking system.

4. **Data Analysis and Storage:**A solution dependent on Elasticsearch and Greenplum is embraced and adopted for big data usage in the freight management domain. It uses the data warehouse of Greenplum having MPP architecture to store organized data. It also applies HDFS conveyed document framework to store huge unique record information and employs open-source web index Elasticsearch to acknowledge capacity and recovery of unstructured and unorganized data. Ordering between various capacity information bases works with the affiliation and review

of information. This stage gives a common library for all the algorithms stored in the system which consist of clustering algorithms, recommendation, and prediction algorithms, association rules and more. It fabricates a specific and normalized library for algorithms to give a specialized premise for creating data applications which are incorporated with adequate interfaces, is capable of data analysis, multi-dimensional detailing and representation.

8.2.5 WORKING

1. **Construction of Data Models for Cargo System:** This is mainly based on the approach of sources and paradigm. The railway knowledge work platform espouses a distributed knowledge warehouse storage design and makes use of superimposed thought to style the information design, which may clarify structure and intensify data quantifiability. Structured knowledge warehouse layering incorporates the knowledge acquisition layer, knowledge integration layer, and knowledge access layer. As long as single knowledge model is constructed, it will not be able to support knowledge analysis necessities. So as to satisfy the business requirements or needs of numerous levels, totally different knowledge modeling ways ought to be hand-picked in step with the features of the amount.

   To attain quick uploading of the information supply, the information collection or procurement layer utilizes the source-based modeling approach, that is, the information or data structure is according to the information supply. Therefore, the information supply is directly loaded while not reprocessing, that significantly enhances the loading efficiency; the data integration layer is intended to attain the mixing of multiple heterogeneous supply data, therefore its knowledge model has steadiness and quantifiability over the time and changes in knowledge sources, employing a paradigm modeling methodology, that is, employing a three-paradigm approach to make knowledge models in step with the business topic division, that is versatile, scalable, normative, promptly intelligible and might support multiple different applications [17].

To attain economical knowledge access, the information access layer usually uses dimensional modeling ways, especially, exploitation of reality and dimension tables to make knowledge models boost knowledge access potency of applications. Because the knowledge access layer is adjusted to target resolution user analysis wants, it's comparatively bigger variability.

2. **Queries of Cargo Data:** It refers to the fact that customers should have a broad range of options as per their requirements and system should be efficient enough to produce the resultant accordingly. The options provided should be flexible enough to be chosen in any possible combination.
3. **Prediction Models:** Selection of those factors that may have a higher influence on the prediction system id done as per the range of the values of factors that influence the characteristics of the features. Verifying the turning points and evolving trends in the fluctuation period is the basic key to find the fluctuations that occur in the cargo volume periodically.

Prodit model is employed for getting the turning point. Being a discrete model for selection, Prodit model is capable of deciding one in 2 alternative options. Variable R in this model is a randomly generated binary sequence with 0 and 1 as its possible values. Outline the likelihood once the railway freight is within the weak period of P(R=1) and once the railway freight is within the expansion or enlargement period, the likelihood is P(R=0) as in (Eqn. (1)).

$$P = \begin{cases} 1, & t\ represents\ weak\ period \\ 0, & t\ represents\ \exp ansion\ period \end{cases} \tag{1}$$

here; 'P' represents types of cargo in time period 't.'

Maximum likelihood technique determines the Prodit model coefficient that forecasts the probability of cargo volume cycle being in the declination stage. Furthermore, due to the fact that the cargo volume is prone to seasonal cyclic changes, it is important to make sure that a linear equation and a trigonometric function is needed for generating the prediction function (Eqn. (2)). Hence, the earlier cycle data is employed to compute the data required for the prediction of subsequent cycle.

$$y_t = \alpha_0 + bt + a_1 \cos\frac{2\pi}{N}t + a_2 \sin\frac{2\pi}{N}t(2) \quad (2)$$

where; 'Y' is the observation value of 't;' and 'N' is the number of season cycles.

Lastly, the model is analyzed and trained with the help of previous data. While calculating the results from the model, error detection and analysis is also done, and parameters are generated using least square method. This decreases the error in the prediction model.

8.3 ADVANTAGES

There have been constant efforts ever since the 1970s to improve the transportation system and make it smarter. Railway system being the most convenient means for land transportation, it was always given special attention for maintenance, improvement, and further advanced development. It was always seen as having a brighter future because it was affordable by a larger population and had reasonable cost. The concerned transportation authorities have been trying to integrate some of the latest technologies such as sensors, advanced data transmission techniques, and improvised controlling techniques into smart railway system. The main aim is to improve the quality of services currently available for the passengers or clients (in case of cargo), while at the same time making it easier for its employees to work too.

Data is obtained in many ways in smart railway cargo system like from position tracking systems, smart cards, smart detection cameras, multiple sensors, etc. This is why big data analytics is hugely needed for obtaining well organized and meaningful data which is effective in real world. Data generated by the railway management system have grown in a gigantic manner over the year due to this development. Big data is the core as it will be needed for applying any smart technology available in the market today. This is also because the traditional systems cannot process such large-scale data and meet the requirements in a timely manner and hence are quite inefficient.

Smart railways system including the freight department can benefit a lot from big data analytics. Some of these benefits are:

- Data generated from all devices and implemented modern technologies can be stored easily using big data.

- It makes the management process seamless as compared to the results and performance by traditional methods.
- Generated data in itself has no particular meaning. It requires to go through a series of different process such as cleaning, classification, and analysis to make it meaningful and turn it into information or knowledge.
- Big data analytics is the only solution for handling such huge volumes of data which is complex and dynamic in nature. It reduces the overhead of storage, management, and analysis at the same time by using many different available platforms like Hadoop, Spark, and Apache. These are capable of solving all the three problems at one place.
- It provides possibilities of new opportunities and advancement.
- It is capable of improving the operational efficiency such as by assisting in railway cargo traffic management. Collecting and analyzing previous as well as real-time data traffic data can improve the traffic flow.
- It can help the managing authorities to understand the driver's driving pattern which can result in better planning such as the one with best pattern and least no of delays can be given the responsibility of the most urgently required cargo or the most expensive one whereas others can be given the normal ones.
- It was helps in finding the most optimal route which requires least amount of time.
- With the help of prediction methods and sensors being integrated, big data analytics can forecast the possibilities of any accident that might occur and generate warnings accordingly.
- It can improve the safely levels and also help in improving the security on the cargo wagons by adequately showing which one is carrying the expensive items (even if it is missed by the officials handling it) and hence necessary measures can be taken.
- In case any unfortunate accident occurs with the goods being transported or the train carrying it has some issues, immediate actions can be taken by tracking it and providing it with the most easily available services by looking into the analyzed data.
- It can improve asset management by detecting pavement degradation problems, ballast aging and many more.

- Maintenance work can be analyzed and timely done reducing the chances of any kind of failures.
- It widens the horizons for development by solving the most basic problem of them all that is data management and analysis.

Big data analytics is the most basic requirement that can benefit any organization including railway cargo management system in numerous ways.

8.4 CHALLENGES

Everything has its some advantages and disadvantages which should not be overseen. Despite of the fact that the implementation of big data in the field of smart railways cargo management system has numerous benefits, it does come with its own set of challenges which cannot be foreseen. Therefore, before moving towards big data analytics, one should pay attention to these problems as well and take decisions accordingly. The following are some of the challenges:

- The foremost step which is data collection is not that easy in the real world as may seem to be. Because of the constant motion of the train and cargo wagons the data might be incorrect, deficient or inconsistent at specific time period or in certain areas.

  One potential approach for handling this type of challenge is advancement in data gathering process which in turn may improve its capacity. IoT sensors are showing recognizable advancement every year. These can be integrated for overcoming these types of problems. Moreover, the appropriation of information catching mechanization to limit manual information section is likewise fundamental for information quality improvement.

- In this period of Internet and digitalization, the most concerning issue is that of data privacy and protection. These are chances of data leakage during transmission or even storage. Most of the data in railway cargo system is not personal but yet it involves customer id. On the off chance that this information is not rigorously secured, individuals who take this information would hurt the proprietor of the information.

Hence, data security is always an issue with most of the new age technologies. To forestall unapproved divulgence of individual private data, governments ought to create total information protection laws which incorporate what information can be distributed, the extent of the information distributing and utilizing, the fundamental standards of information dissemination, information accessibility and different regions. The transportation divisions ought to rigorously control the individual information definition, reinforce the administration of information security certificate, and utilize further developed calculations to improve the information security level.

- Advancement in the field of data storage lacks the infrastructure for storing the volume of data that is generated every day in today's day and age. In smart railway cargo management system, if we look at the data that will be generated from all the sensors, analysis process, different algorithms that will be used, etc., the kind of foundation that is requirement will not be sufficient with the traditional data storage systems.

  Consequently, before implementing big data, planning the most sensible storage system is the foremost task. Cloud services provided by some of the recognizable names in the industry integrated with such infrastructure which is necessary for implementation of big data have proven to be effective and capable enough.

  Furthermore, consolidating insight with capacity is additionally a decent arrangement. Ventures are searching for brilliant administration apparatuses which can give coordinated examination inside the capacity. This empowers them to lead asset checking and utilize capacity framework.

- Data processing might also be a challenge due to the diverse nature of data that is collected from various sources. Although this problem has already been solved as there are specific big data software in the market which are capable of handling real-time data which is not static in nature or does not have a constant format.

➢ The result of the analyzed data entirely dependents upon the consistency, accuracy, reliability, and volume of data that is given as the input. In order to get more input data, it is to collect more data from different interfaces and multiple users. If the quality of input data is not good enough, it will automatically affect the result which in turn will affect the decision-making process. Encouraging automated data capturing systems or use of AI for data verification might prove to be an effective solution to such challenges.

8.5 FUTURE SCOPE

If we look at the history of railway and railway cargo system, we will realize that we have already come quite far considered the smart railway cargo management system. Apart from all the development done so far it is essential to continue evolving and keep the system up-to-date by advancing it time to time with the new trends in the market. This is necessary because the complexity of data and its codependency will keep on growing with time. If we do not bring in constant updates it the system, after a few years this industry will lack far behind in terms of then present technologies.

- Today's multimodal cargo transportation system needs an advanced interface for administrators, innovation suppliers and public substances to move products in an ideal and proficient way. Therefore, it is essential to pay attention towards enhancing the collaboration.
- Development in one field automatically promotes the development in other fields which related to it. For example, advancement in the field of IoT automatically leads to development of sensors and detection devices. Similarly, while aiming to modify the existing system, many new research works will take place. As a result, it will result in broadening the area for research and technology.
- With all the challenges and the need for constant improvement in the future, there will be a need for group of professional dedicated to for this purpose will be highly required as it is not something which can be understood by everyone. Therefore, people who can

understand technical aspects and have a broader vision for seeing the future scope will be needed.

- The recent surveys state that the workforce in that is currently present in railways or cargo management system is not that skilled and qualified to fulfill the growing demands even with the help of technology. As discussed above, the need for trained human resource capable of coping up with technical advancements. This in turn will open more employment opportunities.
- Once high-speed rail for public transportation is established, it can also be used for cargo. These smart railways with high speed will be helpful for interstate import and export of important and essential goods and specially medicines.
- Use of cloud computing along with big data analytics will help the railway cargo system in improving its network and performance both at the same time.
- Improvement in smart railway cargo system will give rise to better shipping options to the customers at reasonable costs according to their needs and budget.
- Studies on the use of drone for railway cargo system or railway system in general are being carried out. It is being said that integration of drones in the system would increase the safety and reliability.
- Advancement in railway freight management system will give rise to new business models.
- Continuous efforts are being made for transforming the whole system into an autonomous system.
- Keeping digital records of the cargo goods while making it available to the customers would decrease the chances of loss and help in tracking it. It will also make it easier for the people handling the whole process of managing the goods as compared to the current method.

8.6 CONCLUSION

Big data refers to exceptionally large volumes of data which increases exponentially and is dynamic in nature. Big data acts as the core requirement of most of the new age trending technologies like ML, IoT, robotics, automated systems and cloud computing. This is because data is generated everywhere and as the complexity and volume of data being produced is continuously increasing.

The traditional database management systems were not capable of handling such of data. By considering the history of big data, we can understand the improvements in the field of data management and analysis. But applying big data is not the end of the story. Implementing big data in itself is a task that should be carried with utmost carefulness and after considering all the requirements of the system. It requires well trained people for working on big data software. Big data broadly classifies data into three major types that is structured data, semi-structured data and unstructured data. All these are classified as per the general data formats. Working with different type of big data requires different techniques. Same method cannot be implemented to all kinds of data.

Big data is a major part of the Fourth Industrial Revolution that is often termed as Industry 4.0. Industry 4.0 embraces the ideology of computerization and digitalization for simplest task. And as this results in producing gigantic amounts of data, big data becomes the most fundamental requirement in the current revolution process. Fourth industrial revolution is driving the companies and organizations towards automation which will further increase the data production rates by considerable amounts.

Today, the concerned authorities are putting constant efforts towards making the railway system and railway-based cargo system smarter by integrated some of the latest technologies in order to get several benefits such as better asset management, improvement in operational efficiency, error detection, accident prediction, customer satisfaction, reducing delays and many more. This will eventually lead to firm usage of big data in smart railway cargo management system because raw data in itself has no meaning and cannot be used directly for any decision-making process. Therefore, data cleaning, analysis, and management are the crucial steps for producing some information as its result which can further act as knowledge after discovering patterns and can assist in decision-making processes.

Big data analytics can largely affect the transport industry as well as the SCM. Moreover, with all the research going on and good future scope it can go further by characterizing new choices and ways to seek after. Maybe then separating an organization to known factors and potential outcomes, huge information in the transportation business is near the precarious edge of making the inconceivable, conceivable. However, the main enhancements are probably going to spin around client care as the entirety of the advantages of huge information mix into a solitary, charming, and commonly helpful connection with clients.

The numerous advantages of smart railways cargo management system come with its own set of challenges which cannot be overseen as ignoring them might affect the resultant data which if used as input for decision-making algorithm or prediction algorithms, will give inadequate results making them highly unreliable and prone to errors and failures. Before moving on to big data analytics for improving performance, organization, and authorities focus on the most basic thing: infrastructure that is capable of storing and operating enormous volume of data. This is crucial because it will be the foundation for all further developments and administrative work.

Implementing big data and making smart railway cargo management system is not the end. There is always scope for more improvement. Smart railway cargo management system has great future scope. Steps towards improving collaboration, conversion to autonomous system, high-speed rail cargo, implementing cloud computing and using sensors for efficient and real-time data collection are some of the aims for development in near future.

KEYWORDS

- **big data**
- **decision-making**
- **industry 4.0**
- **real-time data collection**
- **smart railway cargo management system**
- **traditional data storage systems**

REFERENCES

1. Chun-Wei, T., et al., (2015). Big data analytics: A survey. *Journal of Big Data, 2*(1), 1–32.
2. Gokalp, M. O., et al., (2016). Big data for industry 4.0: A conceptual framework. In: *2016 International Conference on Computational Science and Computational Intelligence (CSCI)*. IEEE.
3. Xu, M., Jeanne, M. D., & Suk, H. K., (2018). The fourth industrial revolution: Opportunities and challenges. *International Journal of Financial Research, 9*(2), 90–95.
4. Philbeck, T., & Nicholas, D., (2018). The fourth industrial revolution. *Journal of International Affairs, 72*(1), 17–22.
5. Obitko, M., & Václav, J., (2015). Big data semantics in industry 4.0. *International Conference on Industrial Applications of Holonic and Multi-Agent Systems*. Springer, Cham.
6. Rühr, A., et al., (2019). A classification of decision automation and delegation in digital investment management systems. *Proceedings of the 52nd Hawaii International Conference on System Sciences*.
7. Glassman, P. A., et al., (2002). Improving recognition of drug interactions: Benefits and barriers to using automated drug alerts. *Medical Care*, 1161–1171.
8. Ghofrani, F., et al., (2018). Recent applications of big data analytics in railway transportation systems: A survey. *Transportation Research Part C: Emerging Technologies, 90*, 226–246.
9. Balog, M., Hanna, S., & Angelina, I., (2019). Intelligent systems in the railway freight management. *International Scientific-Technical Conference Manufacturing*. Springer, Cham.
10. Galar, D., Kumar, U., & Karim, R., (2017). Big data in railway operations and maintenance. *Globe Railway Review, 4*, 11–14.
11. Chen, M., et al., (2018). Smart-rail-smart supply chain-oriented rail freight service. *Proc.: 7th Transport Research Arena*.
12. Zhibo, C., Xiaoning, M., Taifeng, L., & Zhengyang, Z., (2020). Research on freight big data application based on railway data service platform. In: *2020 IEEE International Conference on Artificial Intelligence and Computer Applications (ICAICA)*.
13. Zhu, L., et al., (2018). Big data analytics in intelligent transportation systems: A survey. *IEEE Transactions on Intelligent Transportation Systems, 20*(1), 383–398.
14. Senthilkumar, M., (2018). Energy-aware task scheduling using hybrid firefly-bat (FFABAT) in big data. *Cybern. Inf. Technol., 18*(2), 98–111.
15. Ferreira, L., & Martin, H. M., (1997). Modeling rail track deterioration and maintenance: Current practices and future needs. *Transport Reviews, 17*(3), 207–221.
16. Senthilkumar, M., & Ilango, P., (2016). A survey on job scheduling in big data. *Cybernetics and Information Technologies, 16*(3), 35–51.
17. Senthilkumar, M., & Ilango, P., (2020). Energy aware task scheduling using hybrid firefly-GA in big data. *International Journal of Advanced Intelligence Paradigms, 16*(2), 99–112.

18. Pan, G. W., & B. Faulker. (2002). "Guanxi and business network relationships associated with the Chinese outbound tourism market to Australia." *Proceedings Tourism in Asia: Development, Marketing & Sustainability the Fifth Biennial Conference, May 23. Vol. 25.*

CHAPTER 9

Transport in the New Era Using Artificial Intelligence

RAJBALA,[1] PAWAN KUMAR SINGH NAIN,[1] and AVADHESH KUMAR[2]

[1]*Department of Mechanical Engineering, Galgotias University, Greater Noida, Uttar Pradesh, India*

[2]*School of Computing Science and Engineering, Galgotias University, Greater Noida, Uttar Pradesh, India*

ABSTRACT

The rapid growth of AI technology (AI) offers exceptional options to strengthen the productivity of various industries and businesses, including the automotive industry. AI innovations include compelling computational technologies that mimic how the biological brain functions. Artificial intelligence (AI) in mobility attempts to tackle the challenges of increased travel demands, emission levels, occupational health and safety, and environmental devastation. The wealth of statistical methods and AI in this digitized world has made it considerably more feasible to solve these problems more comprehensively and expensively. Several of the automated processes that are making their entry into motorized vehicles are as follows: convolutional neural networks (CNNs), evolutionary algorithms (GA), evolutionary computation (SA), biological immune system (AIS), ant colony optimizing compiler (ACO) and honey bee optimization (BCO) and proposed fuzzy models (FLM). A thorough grasp of the relationships between AI and collaboration and team on the one extreme, and public transportation characteristics and aspects on the other, is required for effective AI implementation. Furthermore, transportation organizations have the chance to discover how to employ these innovations to give speedy

Smart Mobility and Intelligent Transportation Systems for Commercial and Hazardous Vehicles.
T. Poongodi, Naga Swetha Pasupuleti, and Balamurugan Balusamy (Eds.)

relief from crowded areas, make traveling time more reliable for their clients, and enhance the economies and reliability of their critical assets. This chapter reviews AI technologies used to address difficulty walking throughout the world, focusing on transportation planning, disaster prevention, commuter trains, and transportation systems. The analysis proceeds with an assessment of the problems of transportation-related technologies.

9.1 INTRODUCTION

Machine intelligence (AI) is a vast area of research and development that enables machines to behave similarly to people. It is used to solve problems that are difficult to solve using typical computing models. John McCarthy developed artificial intelligence (AI) in 1956 [1]. However, it struggled to achieve its goals, and a measure of failure resulted. Technological advancements made it less practicable. Scientists researched AI utilizing Experience and understanding systems between 1960 and 1970; KBS and Autonomous Neuro-fuzzy techniques were developed (ANNs) [2]. KBS technologies are programs that give guidance based on pre-existing knowledge supplied by people. ANNs, on the other hand, are networks of neural pathways arranged in layers corresponding to the person's intelligence that have been applied in medicine, biology, linguistics, architecture, law, production, and other industries. Because of the restricted application of ANNs and a lack of information until the 1980s, interest in AI dropped during that timeframe. Since the 1980s, various studies have been undertaken to understand as much as possible using a process known as backpropagation [3]. This approach is known as a Backpropagation algorithm during the learning phase, and it has been used to address problems in various fields with a single convolutional. Nowadays, the abundance of data has given rise to learning algorithms as a subclass of AI. Instead of educating algorithms anything, machine learning (ML) entails programming them to act like human brains. It offers computers accessibility to large amounts of data and extracts significant aspects to solve complex issues [4].

The most well-known AI method, ANN, is used in various industries. The based on neural networks (NNs) has been one of the original and most prevalent forms of ANNs, in which data flows from perceptron to outputting vertices through a specific cable. Neural networks (CNN) and

Reinforcement Learning (RNN) are two different forms of ANNs. CNN specializes in visual identification tasks, but RNN analyzes several inputs very well with a spectrum of uses such as language, composing, and document classification. The various hidden layers interwoven in their design are sometimes known as Better Educational Methodologies [5].

Several ambiguities and holes in the information cannot be resolved using typical methods. As a result, AI makes advantage of these inconsistencies to construct a link between both the incidence and impact of various real-world occurrences by combining current data with assumptions and alternatives for a much more precise evaluation. Transport problems arise when the architecture and consumer engagement are too complicated and predictable [6]. As a result, AI can be seen as an ideal fit for mobility infrastructure to tackle the difficulties of rising tourist spending, Global warming potential, safety problems, and destruction of the environment. These issues occur due to the continual increase in private and public traffic caused by the growing population, particularly in developing nations. The cost of overcrowding in Queensland will be 53.3 trillion dollars by 2031, as the population rises to 30 million. More than 640 km of main highways in Hobart, Australia, are congested during peak times, culminating in 2.9 tons of Carbon dioxide emissions per year [7]. Many academicians in the 21st millennium aim to develop a more trustworthy mobility network that has a lower impact on people and the environment by applying expenditure and much more trustworthy AI. It might benefit mobility infrastructure, drivers, vehicles on the road, and autos.

Logistics AI technologies have grown and been deployed in a variety of ways. Among these, this research aims to look into three important examples. Machine intelligence (AI) is used in business decision-making, advertising, and administration [8]. This is crucial in addressing the issue of continually increasing demand in the presence of significant route accessibility. This entails using correct diagnosis and classification techniques to anticipate the movement of vehicles, driving conditions, and events. AI technologies that enhance mass mobility are also discussed. It is because the transportation system is seen as a more environmentally friendly method of transport. (iii) The subsequent intriguing AI use in mobility is networked and autonomous vehicles (AVs), which attempt to boost output by decreasing waste. This chapter discusses the self-driving automobile and innovative connected bus trials launched, most notably in Sweden, Thailand, and Chinese [9].

9.2 ARTIFICIAL INTELLIGENCE (AI) USE IN TRANSPORT NETWORK

In several instances, it is hard to completely comprehend the relationships among the various parts of the transportation system; consequently, AI approaches may be provided as a reasonable answer for such complicated structures that cannot be controlled utilizing standard processes [10]. Several studies have shown the benefits of machine intelligence in transportation. An example is transforming road traffic surveillance into a helpful tool that recognizes incidents and anticipates future traffic circumstances. Furthermore, many AI methods, such as ANNs, are being used in mobility. ANNs may be used to drive directions, identify public transit traffic incidents, and predict traffic [11]. It is classified into unsupervised teaching methodologies and unsupervised learning techniques. Optimization learning includes machine learning approach (SVM), probabilistic neural network (PNN), Nonlinear Activation Network (RBN), K-Nearest Neighbors, Linear Regression, and unstructured NNs, including opportunism layer-wise and principal component analysis. Several transportation difficulties result in an easy solution that necessitates the development of custom algorithms to make quantitative analysis simple to solve [12]. They are raster algorithms, which are very advanced computer techniques. One of these techniques is the genetic algorithm (GA). It is based on the natural concept of evolution. It handles complicated optimization issues using the "survival of the fittest" concept, and it is a valuable instrument in urban planning networking. Search algorithm (SA) seems to be another technique that simulates the metal roasting method [13]. ant colony optimizing compiler (ACO) is another AI system developed by people based on the actions of a swarm of live ants as they travel from the nests to food availability. A computer-based autonomous system protects the body (AIS) on the host immune response. Bee colony optimization (BCO) is a method for resolving a hybrid significant-scale optimization issue. Computational intelligence systems include ACO and BCO [14]. Swarm intelligence is an adaptive mechanism motivated by ants and bees operating together in the same swarm to find the best answer. Because these systems' computational intelligence analyzes can represent unpredictability, imprecision, and ambiguous notions, these approaches are helpful for path optimization challenges in transportation. The fuzzy logic model (FLM) is another optimization tool (FLM). It is used to address the shortest motion planning problems. When building an optimal route model, the effectiveness of FLM is compared to that of the logistic regression model (LRM), and FLM outperforms. As a result, intelligent approaches like FLM, GA, ANN, and ACO are well suited for predictions, thinking, and adaptation. As a result, these are employed to solve optimization issues involving actual-time traffic circumstances and occurrences. Agent-based software engineering (ABSE) is a new software methodology based on AI ideas that have been developed (ABSE). ABSE can provide a nonlinear mechanical for discovering the shortest distance by combining several criteria and circumstances [15].

Artificial neural networks (ANNs) integrated the system with the approaches above for improved outcomes. For example, citations supplied another AI tool utilized as an operation architecture project for automated autos and travel routing. Transportation operators must identify when and how to use these innovations to function, decrease congestion, boost travel time consistency for their customers, and improve the profitability and effectiveness of their vital infrastructure [16].

9.2.1 ARTIFICIAL INTELLIGENCE (AI) IN TRANSPORTATION INFRASTRUCTURE STRUCTURAL DEVELOPMENT, DESIGN, AND ADMINISTRATION

The purpose of transportation development is to evaluate the needs of local communities and decide how to address those needs while taking into account the sociological, environmental, and socioeconomic disparities [17]. The network design problem (NDP) involves an ideal road method for transportation planning (NDP). When the capacity of current additional infrastructure (expand link utilization, intermediate, and shoulders area), a Discrete difficulty is recognized, and a Mixed of Categorical and Continuous problems are identified. Previous studies in the 1990s concentrated on NNs for route planning, engineering, and modeling [18]. For example, allusion used a simultaneous neural network system to predict the spatial link between development and land planning. Following that, research shifted to raster techniques recommended for urban planning since they do not rely on existing linkages and vertices to discover the best path. Today, most research focuses on generating massive amounts of data with powerful algorithms to create patterns within the data using machine-learning. Relate handled the ongoing NDP issue as a Bi mixed modeling problem with two different levels [19]. They tested the effectiveness of GA and SA techniques on a simulated system. SA picks the best bargain with less computation complexity than GA when slow growth. However, GA can yield a more optimal solution because it does more simulations. This conclusion undermines the authors' claim that GA outperforms SA while needing less processing time. On the other hand, the concepts investigated in this chapter were a solitary linear model for a Continuous NDP issue [20].

Another research used computer modeling to create a safe strategic approach for the Turkish city of Ankara. Genetic and ANN algorithms The Training algorithm proved more effective than the GA method, with even more minor inaccuracy. While designing an ideal automobile path, the ant colony method was applied. Furthermore, the reference integrated Cellular Computation (CA), geographical data analysis results, and the Population-based optimizing approach were combined (ACO) [21]. For several years in China, the results revealed an improvement in urbanization number of objectives on modeled industrial land-use changes. Furthermore, the recommendation focuses on developing a solid transportation system health and security strategy to balance traffic demands with suitable highway, railroad, and airways deployment [22]. The author utilized ANN to forecast accidents and injuries on a heavily used road from Istanbul to Ankara. Finally, in Poland, reference was evaluated using GA, SA, and AIS approaches. The best modification for conventional network architecture is discovered while seeing it as a Mix NDP issue. The results demonstrate that SA beats GA and AIS for this problem [23].

Additionally, a vehicle mobility system is required to alleviate traffic congestion and travel time delays. Many scholars concluded that the ant colony method might be used to solve the problem of traffic congestion. At the same time, they are focusing on using the BCO Algorithm to solve a route planning problem [24]. The task requires choosing a path inside a network and designating a wavelength to that path for each associated node to calculate the quantity of directly connected networking devices. Furthermore, recent research has focused on using nanoscale traffic data to model and discover security problems, traffic control systems, and highway plan development. The reference advises learning and updating the real-time path creation system based on traveler preference to determine the optimal routes for public transportation customers. In addition, a utility-based strategy that focuses on distinct qualities of pathways and parameters for each public transportation customer should be used [25].

Intelligent transportation systems (ITS) is another area where AI technologies have experienced tremendous progress. Using various technologies and telecommunication networks, these systems attempt to reduce traffic congestion and increase driving experiences. They collect critical data that may be used with ML technologies. Deep structured learning or hierarchical knowledge, for example, has been used to optimize transportation control principles in massive ITS systems in the context. Similarly, a fully convolutional method has been proposed to enable digital signals and quick response to ITS sensors. Computation analytics functionalities. As it continues to expand, data complexity will rise, necessitating DL techniques to uncover patterns and characteristics in this data to create more integrated transport networks. A ML model and fuzzy approaches were utilized to manage traffic signal installations at crossings autonomously. The proposed ITS system, 'NeverStop,' used RFID detectors, and the outcomes demonstrated that it successfully decreased typical vehicle long waits. In addition, two NNs systems that are based on little numerical simulations were created to improve road regulation. The first technology is also in charge of traffic signal management, while the second is forecasting future congestion problems. At the same time, it is creating a multi-layer NNs technology that was tested on providing a better understanding system to demonstrate the practicality of using NNs to govern traffic [26].

ANNs are also useful in signal congestion control. Two NNs systems have been designed to regulate the highway more efficiently based on microscopic simulation results. The first software automates street lights, while the secondary anticipates future traffic problems. At the same time, the Government proved the viability of employing NNs to regulate transportation by delivering a multi-layer NNs technique that was examined to have a different knowledge of circuitry [27].

Additionally, when the program's characteristics and cycle time are updated, data gathering NNs are used flow changed regularly. AI is an active study topic that is constantly evolving, and new approaches and applications are routinely launched to leverage the power of AI to enhance road construction, decision-making process, and maintenance.

9.2.1.1 *REDUCING CONGESTION*

Many individuals have determined how long, place, and intensity of an occurrence and help congestion. Management in reducing congestion. Manual reporting, automatic algorithms, and Bayesian networks are examples of these endeavors. Automated documents produced by humans may delay recognizing events and be less cost-effective [28]. On the other hand, Computers can evaluate the aerodynamic performance before and after the event using data acquired from monitors along the route. Parameters can be determined, such as the California Algorithm, which is used initially to construct the event detection method. However, it is impossible to deploy an algorithm on arterial routes due to street parking space signals. As a result, neural network techniques have resulted in the development of algorithms. A categorization neural network method was explored to detect incident occurrences on a motorway [29]. Another study looked at developing a suitable software suite to recognize all vehicle objects in the scene. For accurate picture recognition, the author recommends utilizing AdaBoost software. To detect occurrences in composite state situations, the precedent presented the IMM ENKF algorithm. Then, a more effective method called "efficient multiple model particle filter (EMMPF)" was created to recognize accidents on the roadway using both simulation and field data. Furthermore, as described in, occurrences can be noticed in real-time via social networking sites. It shows that Twitter is an expensive and convenient approach for recognizing incident occurrences on the motorway and peripheral roadways [30].

9.2.1.2 *FORECASTING FRAMEWORK*

The fast growth of interconnected transportation infrastructure (ITS) has raised the requirement for better techniques for forecasting incoming traffic. These methodologies are critical to the success of ITS components such as enhanced traveler computer systems, sophisticated traffic control systems, improved public transit systems, and passenger car capabilities. Traditionally, data collected from highway sensors have been utilized to build intelligence guidance documents. The knowledge is then incorporated into computational intelligence algorithms to provide real-time, short-term, and long-term projections [31].

Previously, research concentrated on short-term flow predictions using a primary feed-forward neural network. An artificial neural approach with one hidden state was incorporated into the entire urban traffic light control system by reference. Using solely simulation results, the researcher proved the high capability of forecasting traffic flow for up to 1 minute. In contrast, the precedent used data sets from such a 1.5 km stretch of roadway in Brisbane, Australia. A time-lag regenerative network was used to create an instrument neural network architecture (TLRN). With 90–94% accuracy, the model predicted intensity for the next 5 minutes. Furthermore, velocity and flow were employed as network inputs; they indicated travel time with the same precision as performance forecasts [32].

In another scenario, the researchers utilized a DL method to predict pedestrian movement up to 60 minutes ahead of time. Data on traffic conditions were acquired from Texas highways. The team used an invasive layer-wise technique to conduct an unattended stacking of vehicle (SAE) modeling. Because each output is sent back into the human brain, it captures crucial properties of signalized intersections. Then, for prediction, an unsupervised multinomial logit layer is used [33].

The system outperformed previous computational methods for nonlinear spatially and temporally dimensions traffic data association, SVM, for example, and Multi-layer NN/Back NN was developed and maintained. Nonetheless, the intended unconstrained probabilistic neural networks (DBN) was trained to utilize a cautious layer-wise approach to learn critical different flow properties. Then, for forecasting, an unsupervised regression layer was used. The author suggested a multifunctional regression layer using a weight aggregation approach to connect numerous jobs and train the model. The model's reliability improved as a result of this. Furthermore, they created a deep training algorithm utilizing field information recorded in California to forecast the protracted flow of traffic [34]. The models successfully gathered topographical existing traffic data from the DL model and chronology traffic flow features from the back-propagation technique. Another artificially intelligent (AI) method to vehicle detection is presented by. The creators built a rudimentary multi-layer perceptron for relatively brief prediction using Wade's mathematical model. It differs from standard feedforward NNs due to the addition of a background layer. This background layer serves as a jewelry box, retaining past information. The evidence received at (t–1) is then transmitted back to the concealed state one at a time, along with the inputs (t). This helps the network forecast the substring; as a result, it is sometimes alluded to as "Jackson's Sequence Networks." The incoming data consisted of vehicles from Italy's road infrastructure management system. As a result, the anticipated traffic flow in the future. This network may be taught as a fodder convolution layer to decrease error using a lumbar instructional strategy. However, the researchers discovered that the networks performed best, when possible, units would be double the quantity of nodes in the input layer. Flow performance is estimated to be somewhere between 92% and 98%, given a 0.5 proportional gain and an increasing muscular endurance. Since this data is designed a first order management system, which provides incorrect projections whenever higher cognitive kinetics are calculated, the standard advocated that the network be found that this approach at each positive consistent online to ease algorithms computation. Another type of sequential circuit is the Elman Connection. The ATOS Scientific Establishment also created a pattern-based framework (PBS). The first categorization job is a training collection containing annotated data and an acceptable output. Second, there is an ensemble classifier, in which model parameters are used to identify trends and pick the suitable work; and moderately continuing to gain knowledge. Recognized data is blended with a significant handful of larger datasets for modern computing research. The city of Berlin presented details for movements and velocities distributions. The findings demonstrated an angular momentum accuracy rate of 93% and gas pressure prediction performance of 86%. Figure 9.1 depicts three components that must be recognized to create a robust prediction model.

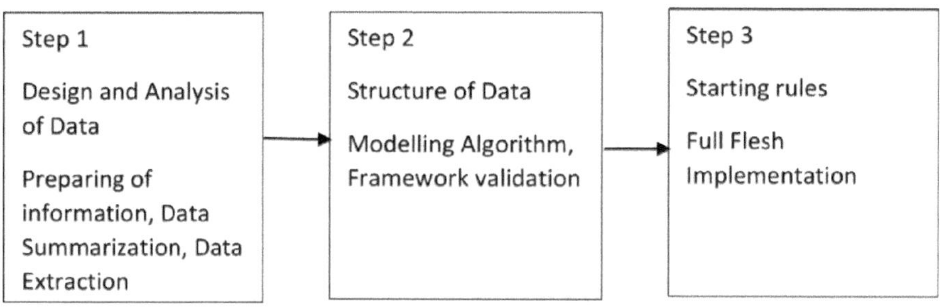

FIGURE 9.1 Forecasting framework.

All information systems must constantly be reviewed and utilized to incorporate the relatively powerful into the company within the first stage. This stage is crucial for the baseline examination of resource use. Several designs can be chosen for a more comprehensive functionality appraisal process during the active experimental phase by defining Breakdown mechanisms. The period necessary over the project's whole regular living Furthermore, the graphic's final portion represents real-time property standard evaluation information since modeling at this point should be frequently maintained and altered to deliver the most outstanding results [35].

Using demographical data, the scientists project innovative mobility demands in Austria. This requirements evaluation is required to develop scheduled and strategic orientations that will most effectively operate a more economic mobility network. They start by classifying the aggregate characteristics depending on individual commuting choices. The data is then classified using a prediction model, and a ML model is utilized to retrieve more essential attributes from the data. They are both supervised NNs. A clustering algorithm employs many de-correlated decision trees to form a forest. The greater the number of trees, the greater the precision. It operates by Each tree (x) is classified based on its attributes. The trees voted for (x), and the forest chose the most points category. On the other extreme, SVM classifications output by minimizing disparities between data points. Then, using the ML approach, they forecast future utilization based on a vehicle's daily miles traveled. Another research looks at options for concise predictions of public transit impatient customers at bus stops. Stations as well as onboard customers. This data will assist operators in better controlling transport journeys. It also helps travelers determine the best course of action during crowded times. Information and communications necessitate merging public transportation systems, improving predictive modeling for processes, and then continuing information management and operating control [36].

Similarly, new transportation companies like Uber have increased the ability to capture large volumes of information. AI can utilize this information to forecast transportation capacity accurately, avoid empty busses and lower congestion and energy consumption. The Multi-View, Geographically, and Temporally (DMVST) Network is used to create an unsupervised learning approach. The researchers acquired large-scale traffic information. They merged Localized CNN, which records locations regarding their surroundings, with a long–short-term learning approach (LSTM), reflecting temporal dependencies. The results revealed that the proposed model outperforms others. Likewise, in Tokyo, Japan, a Multi-layer Fourier transform was Developed. The knowledge was utilized to anticipate taxi demands. They collect data using the Taxi Probe program, which equips cabs with personal information devices (e.g., where the taxi is available). According to findings, employing 4 hours of previous customer information with 50 nodes in the hidden layer enhanced predictive performance. Similarly, a related study employed L1-Norm SVM to measure taxi efficacy by selecting the essential characteristic from a taxicab sequence. Another research used data mining algorithms to forecast taxi trip requests in New York City, USA. Although the DNN beats prior AI systems, a suitable design should be identified to achieve promising findings.

Ultimately, AI can play a significant role in avoiding and minimizing the effects of urbanization car accidents. Automotive mishaps are caused by various factors that vary in place and time. Consequently, AI may mine information for geographical features of incidents and phases that can be exploited to develop protective measures. For example, a study in Beijing, China, used a deep neural method for predicting the probability of traffic deaths by analyzing temporally and geographical information from a head-on collision dataset. The results revealed that this method is efficient and might warn people about potentially harmful circumstances. Other investigations created a technique known as Stack Change in status. Convolutional Neural Computation to predict the severity of car accidents. Another research finds that computing approaches (k-Clusters and the primary method) can extract crucial characteristics from database created vehicle collisions. Thus according, the most important factors connected with the potentially deadly frequency of an event that happened in Abu Dhabi were: sexual identity (male with the required level of traffic fatalities), age (among both 18–30 years old), crumple zones type (car to accessibility coming into contact has been the most prevalent in crashes), as well as the position of an incident that occurred (right angles). Computer science approaches such as Regression Analysis, and MLP were used to analyze these characteristics [37].

9.2.2 IMPLEMENTATION OF ARTIFICIAL INTELLIGENCE (AI) IN AERONAUTICS AND PUBLIC TRANSIT

Agricultural data collection has become much more trustworthy and controllable as quantitative computing has advanced. Although urban transport data is large and disorganized, it should be appropriately documented to profit from it entirely.

9.2.2.1 FLIGHT CONTROL

ML has now been praised for its potential to promote flying control. AI can aid in the supervised learning of inventions, operational systems design, and software development (Intelligent Maintenance, Flight Route Optimization). The (PLADS) technology was designed to obtain very density flight information and make vector propagation easier (SVR and SA algorithmic platforms). It was demonstrated that SVM produces adequate findings for this form of categorization. In contrast, precedent showed that the unregulated learning algorithm helps boost landings assurance when an aircraft is landing. For example, references assessed the plane's security by analyzing the onboard engine using the DL architecture (PNN). PNN design and simulation are identical to MLP learning, except that a feature representation is employed rather than a linear function, and the RBN concealed layer requires many more connections. Furthermore, they created an autonomous unsupervised random forest (RF) method to identify aircraft turbulence better. This device can assist the pilot in avoiding deviations from the predefined path, reducing fuel costs, and improving flight control administration. Finally, the reference demonstrated the need of employing ANN to improve the monitoring program of the gas turbine during flying. To get the maximum precision needed to improve the monitoring program, the author advises assigning each element in the framework to a diagnostic NN [38].

9.2.2.2 SHARED MOBILITY

An economic system in commuting has become a promise, especially in dense metropolitan areas. It has resulted in reduced inner-city congested roads, with the possibility to provide alternatives to eradicate solitary or less occupant individual autos from the highway. Consequently, this allows for a win-win approach that provides both short-term socioeconomic benefits and proper long-term population via extra adoptive families to reduce traffic difficulties and environmental ramifications. The collaboration economy can be characterized broadly as follows.

"ICT-enabled platforms for exchanges of goods and services drawing on non-market logics such as sharing, lending, gifting, and swapping as well as market logic; renting and selling."

The digital economy is gaining traction primarily in the automotive and lodging sectors, with Uber and Airbnb leading. In the transportation industry, the burgeoning digital economy may have a prospective interest in establishing business strategies to generate new-shared mobile financial services to develop a stable transportation network and more efficiently manage the deficit of matching demands to supplies. The bulk of consumers' attention has been drawn specifically to on-demand renting of bicycles, automobiles, and transportation. These public transport solutions also solve the primary flaw of "first-mile, last-mile" transportation [39].

Transportation systems and vehicle swapping are examples of municipal transportation ideas that have been present for generations; in fact, the first scooter systems debuted in the Netherlands in the 1960s. However, as digitalization have advanced, current improvements and innovations of disrupting ideas in public transportation have been feasible on a large scale. Furthermore, these app-based systems coordinate on-demand automobiles and ride-sharing for a long-term sustainable transportation system.

AI system enhances there have firms by improving consumer satisfaction and optimizing operations. It is now possible to give users personalized customer interactions by combining AI with mass transportation. Uber, for example, has built a personalized bed travel experience proposing the destination of riders are determined by the user's prior excursions. Furthermore, Uber has adopted stretch of road invoicing, which employs an algorithm to anticipate how much passengers are willing to pay characteristics such as destination, time of day, and geography. Furthermore, the Uber administration is employing machine intelligence to avoid fraud. And detect dishonest drivers. AVs that use AI and sensors technology are envisioned as the future of shared transportation. Automated cars are far more lucrative, and they might replace 6.2 million drivers globally by 2030. AI breakthroughs in transportation have also enabled shared transportation companies such as Uber to plan routes more effectively and precisely predict journey times, encompassing times for pick-up and drop-off. This enhances the stability and improvement of government transportation. Additionally, with the availability of an enormous dataset and Neural, it is now possible to forecast public transport usability of the system based on past journey data and ambient conditions. When vehicles are inactive, an algorithm, i.e., a trip recommendations mechanism, is employed to target potential customers during the registration process. Using DL capacity, Q-learning produced a location optimization technique [40].

9.2.2.3 ROAD TRANSPORT

Even though busses serve a vital part in roads and bridges, several research has been conducted to enhance the security and dependability of bus services and destinations. The Hybridized Ant Colony Algorithms, on the other hand, is employed to control bus timetables. In terms of cab driver scheduling, they are both cost-effective and dependable. By projecting bus flight schedules, ANNs may also reduce passenger wait times. The application of nanotechnology in busses has also led to the development of self-driving busses. iBus was founded with a powerful double modality technology. It contains three primary components. Methods similar to a human operator: monitoring, decision-making, and execution, but it employs hardware and software. In China, this was put itself through the paces. Another form of the autonomous bus is marketed as an Alpha Bus in China. Furthermore, in 2015, an agreement was reached to trial an autonomous shuttle bus at Sentosa, Indonesia. Customers in Malaysia will ride the Intelligence bus in three different locations: Demonstration, Permatang, and the Singapore Technological District, as stated by and by 2022 (JID). In addition, a self-driving Olli bus appeared in the U. S., with tasks ranging from transporting people to their destinations to talking with customers about the journey, the situational characteristics, and Olli himself.

Another feature of AI developments in request-response public transportation is adaptable on-demand regular busses that run on variable schedules and paths. These maintain the desired level to provide entrance connectivity of taxis at a fraction of the cost and the effectiveness of the bus system. Many initiatives and trials are being conducted worldwide to increase the efficiency of traditional bus services and give consumers a more convenient and efficient. Optibus has unveiled an AI-powered On-Time proved helpful. To help transport operators cut delays and provide on-time operation. Beijing has launched a cutting-edge, personalized, and customer public transportation service known as customized bus (CB). In Boston, in 2014, the autonomous flexibility bus service BRIDJ began in

Massachusetts. This uses real-time transportation large datasets and user feedback to estimate how and where people want to go. This program may locate the shortest route and only stops at sites requested by passengers to optimize the service.

It is also crucial to obtain economic viability by selecting the appropriate fleet size and operating at the optimal utilization. As a result, a passenger sales forecast is critical in detecting significant bus progress. A naval size that ensures the most use of space available as a consequence, the outputs of a consumer preferences projection models as a first step, a standard bus network might be used.

Another area where AI technologies have proven their worth is movement security applications. An Automated Traffic Location (AVL) system was employed to improve public transit operational efficiency, handle direct monitoring, and boost overall service quality. This system collects data from the signal transmission to track transportation units instantaneously, identify the challenges to inform vehicles of modifications and plan different routes. It also supplies tourists with knowledge via Applications on their cellular telephones. The materials present an initial decisions movement study of an AVL-based Satellite navigation system was developed in the Region of Italy of Cagliari. Every 30 minutes, the AVL centralized server obtains concrete data on the position of busses. These data are utilized for three primary purposes: improving the integrity of transportation procedures, prioritizing bus activities at sidewalk demarcation intersections, and informing travelers about bus timetables at benches.

The bus system in London, as earlier indicated, is also another illustration of an AVL-based Navigation system. Furthermore, a novel AI-based information management system has been presented to assure the smooth execution of AVL systems. So according, pattern recognition clustering techniques were used in conjunction With AVL data to improve the AVL company's capacity for arranging public transport in Portuguese. This interconnection helps analyze if a schedule adjustment would meet the cable network needs. In that other case, the authors proposed employing uncontrolled clustering computational methods to discover trends in AVL data to optimize public transportation timetables in Sweden. Furthermore, specialists employ AVL information in BusGrid technologies to anticipate future transportation demand on public transport and passageways using data mining algorithms to improve the creation of significant routes and bus schedules. According to the research, the system can adequately anticipate constraints of the system at bus shelters. However, as noted previously, there are significant challenges and areas for further study in AVL-based public transit planning. First, there is a shortage of evaluation of predictability for current scheduling planning (SP) signals by analyzing periodicity in each pathway. Furthermore, determine if SP can cover the behavior of modern systems and fulfill the present customer preferences. The second distinction is the examination of employing ANN to improve journey time prognostication logistic issues and to improve the location of the bus time schedules using logistic regression and bandwidth The Kalman filter. The final goal is to develop an industrial automation strategy focused on AVL and Autonomous Passenger count APC information. For rapid event reaction and optimal public transportation scheduling. In another paper, the authors discuss the significance of using contactless payment data to predict and predict the demand and behavior of public transportation users. Because most intelligent card data only offer entrance data for passengers, several studies have been undertaken to estimate the destinations of people on the plane using various approaches based on this data. Reinforcement learning is one of these methods used. Techniques are described based on intelligent card information accuracy, transportation mechanism, statistical significance, information sources, confirmation (algorithm quality), situation planning, and parameters influencing the recommended techniques. The authors discovered that most publications do not evaluate their simulations and do not conduct risk assessments. The researchers also suggested that the current models be enhanced by incorporating land-use issues and information regarding public transit infrastructures. Many computational modeling, such as AVL, video recording, and GPS data, may be evaluated and integrated to anticipate the patient's location.

9.2.3 SMART URBAN TRANSPORTATION

Intelligent city transportation's future ambitions include thoughtful decision examples from natural intelligence and system integration through optimum equipment usage. It builds a more robust, safer, and healthier mass transportation system that promotes cognitive connectivity to generate long-lasting, frictionless, and ecologically responsible infrastructures. The world was previously on the edge of automated driving (AV), or automobiles that can operate without the help and monitoring of a single operator.

9.2.4 ARTIFICIAL AUTOMATED VEHICLES

"Automated driving (AV) relies on machine intelligence (AI) software based on AI techniques. This system works by teaching the car how to drive while maintaining safe progression, change lanes, and management," among some other things. AVs are predicted to have a considerable impact on how significant highways are handled worldwide, and their implications on traffic control and congestion, as well as their capacity to affect tourist behavior, have been predicted in some detail. Due to these AVs, people's travel patterns are expected to change, giving rise to new governmental systems and urban structures. They will contribute to car switching and ride-sharing by creating new business tactics to meet existing restrictions such as limited stability and performance.

The authors explored the benefits of identity automobiles. It was initially envisaged in the late 1950s, and that became a reality in 1968 with the introduction of the "urbmobile." The Cornell Aeronautical Laboratory created the first petrol vehicle, but current technology did not encourage its adoption. The rapid development of surveillance systems has highlighted that it is necessary to develop more dependable and safer roadways by entirely automating autos for the intended destiny of transportation. On the other hand, other experts feel that it is hard for the automobile to recognize its surroundings in metropolitan areas since it requires a large quantity of data to detect traffic lights, other vehicles, yellow lines, and so on, especially in poor regions lighting situations. Furthermore, computer problems like confidentiality and dependability are essential concerns for these technological developments.

These self-driving cars are mainly composed of two components: equipment/software framework. The equipment framework includes controllers, detectors, and a programming language, while the programming consists of navigating parts and an operational framework. Fingerprint technologies and networking between connected autos support a localization method and a skeletal receptor for recognizing movement vehicles AVs. The next stage in the evolution of this AV is fully automated cars, with the challenge being to enable the automobiles to securely operate in between some of today's trucks on the road while disregarding obstacles and people. Knowing deep intelligence, computer processing technologies, and the use of sensing and 3D webcams can help you develop this expertise.

Google unveiled their Toyota Prius autonomous vehicle in the United States in 2010. Furthermore, because the car would park itself in isolated regions, the demand for parking spaces will be reduced. Google released the Lexus RX450h in 2012, and in 2015, a completely autonomous car with no steering wheel was launched under the moniker "Firefly." According to, Google is now working on their driverless car project through a separate firm called "Waymo."

However, as mentioned, societal acceptability is vital to recognize as a barrier to customer adoption. Global research was done on a separate component of making the cognitively intelligent car more enjoyable to travelers. These features were as follows:

1. **Self-Healing:** Automobiles can detect and correct mistakes on their own.
2. **Self-Socialization:** The ability of an automobile to converse with its environment, other cars, and humans using speech recognition.
3. **Self-Learning:** The vehicle understands its behavior and the driver's, customers,' and surrounding components.'
4. **Self-Driving:** A vehicle's ability to run autonomously in a controlled area, with some automated constraints.
5. **Self-Configuration:** Each transportation device has digitized information to achieve the overall and distinctive vehicle sensations.

Notwithstanding the difficulties, there are several advantages to adopting self-driving vehicles in the future, including more excellent accident prevention, less congested and delayed, faster commute times, fewer automotive repairs, lower emissions, and improved fuel mileage. Because of these benefits, automation transport is an outlay and trustworthy commuting choice in the contemporary.

9.3 DOWNSIDES OF ARTIFICIAL INTELLIGENCE (AI)

AI approaches have sparked particular outrage because they have been introduced into the transportation system. One of the most significant barriers to AI is viewing ANNs as a "metal box." This way, the relationship between the input and output is developed without knowing the computer's internal calculations. Furthermore, the capability of ANNs to extrapolate in circumstances whereby some data is missing inside this large dataset became questioned. However, research has overcome this obstacle by merging NNs with various traditional methods and other AI tools as a hybrid approach to resolving this issue. However, the current drawback is the demand for hybridized to improve comprehensive effectiveness, particularly in multi-situations. Because of the introduction of a mechanical system, developing an AI-primarily dependent completely for a sustainable transportation device may be difficult. Understanding and correcting facts made complete data Until now, it might be more environmentally friendly if AI modules could manage the entire process. As a result, understanding the full potential of AI to expand modules capable of functioning as separate components are required. As a result, in future studies, it's much vital to use AI know-how in visiting assessment, data processing and preservation, decision making, and optimizing modeling. When AI tactics are primarily focused on data acquired via traditional approaches such as the employment of different sensors, cameras, actuation, etc., the timeliness and completeness of forecasts aren't dependable. As a result, transitioning from traditional data collection approaches to new AI-based analytics technologies is crucial, as it may provide unique and innovative data mining tools. Another challenge is easily transportable. Another challenge is obtaining the best gold solution of known concentration for AI technology's graphics techniques (e.g., GA and ACO). Furthermore, in opposition to the employment of AI raster procedures, quantitative computational approaches can provide accurate knowledge on the underlying geometry of the problem and the characteristics of the solution. However, in authoritarian optimizing methods where appropriate mathematical procedures cannot be used, the immediate analytical impacts provided by those techniques outperform no solutions. Furthermore, the experiments mentioned above showed that those techniques were satisfactory in most cases. The settings and preconceptions must be changed and repeated several times to obtain good results and deeper insights into the problem. Another issue is the injustice built into college transcripts, often proved by persons who can also supply new information. Labeling errors and prejudices project's goal is to foresee unexpected activity and hazardous weather conditions under the surface. Consequently, current AI tactics cannot deal with such behaviors and situations. As a result, improving environment and event response techniques and forecast systems is crucial for achieving high accuracy. In developing these algorithms, AI might influence the productivity of online calculations and standardize the requirements in terms of geographical and temporal data coverage. Most AI techniques that incorporate NN in time collecting distribution products do not frequently contain any testing of the houses of the error and versions specifications, which is seen as a challenge. As a result, knowledge and AI must have an essential connection to progress for history investigations, notably in terms of center simulation and optimization and quality of fit and the capability of effectively and technique for determining massive amounts of data. Aside from the above-mentioned specific AI bounds, a few more fashionable AI lacks.

9.3.1 AI ALGORITHMS

A set of regulations is a rigid collection of principles that focuses on solving a specific problem. Encryption algorithm and assessment are essential components of AI methods. It must be efficient in the time necessary to process the regulations for a large amount of input data. Furthermore, little sufficient utilization of notebook remembrance will determine the effectiveness of a system of regulations. Big O annotation is the most generally used way for assessing a set of regulations. It sets the maximum range of steps that functionality can create. As more data is supplied, the quantity of processes varies; this proportion will assist in establishing how long the method will take to finish the characteristic, i.e., it eliminates all of the quality's elements while the quantitative entry length distance rises. In an ideal environment, as the number of information rises, searches remain unchanged, but the entry duration increases. This, however, cannot happen in a real-life situation. In a basic set of instructions, the volume of entering is surely recommended at a steady rate, while the number of operations increases at the same rate. In distinction, in hard accumulation of regulations scenarios, the number of projects and computations are defined in the mathematical language (X2), exponential time (2X), and exponential time. The intricacy of a system of regulations is represented as a P issue and an NP difficulty in the philosophical method. These issues need quick calculating computer networks to tackle expanding knowledge accumulation complications. P problem can be authenticated by a series of necessities that execute in computation time and are handled with Predetermined Turing Machine, as linked to by methods of difficulties involved that required quick design and implementation. Still, the method for calculating isn't as confusing and hard for complicated sample units as NP. The complexity of the algorithm approaches are used to evaluate NP, and a Non-Deterministic Rotating Mechanism is used to finish the problem. The aim is to have numerous computer networks that provide all possible answers to the same issue simultaneously and provide the optimal answer in a quasi-manner. NP is divided into two types: NP-tough and NP-whole. The NP-tough challenge develops while seeking to broaden instructions that work for each entry. However, NP-complete denotes a problem that cannot be handled in computational complexity and is challenging to provide an approximation answer within the best solution. It encompasses truck routing and efficient software for operations and road users in logistics. The complexity of computation limits AI approaches since most Advanced technologies are characterized as NP trouble or NP-whole complexity. Several successive layers of minerals were discovered on the interior of the population demographics following a thorough analysis. As a result, the challenge stems from the fact that vast volumes of data include distortion and anomalies that may be more difficult to eliminate. In mobility, knowledge may be acquired from various sources, including roadside detectors, thanks to a number, toll guardrails, GPS, Cloud applications, and so on. Massive volumes of data about multiple transport activities are stored in these re-assets. Glide, velocity, and endurance of visitors and passenger behavior are examples of these attributes. As a result, it complicated the calculating method for addressing a certain difficulty. Map-lessen, part of the Hadoop Platform, is one way to perform this operation. To assist in interpreting large quantities of information, the computations are spread within the infrastructure into several machines. The concept was first conceived. The initial education data set is broken into several subgroups and disseminated over overlapping networked computers. Each laptop will independently conduct a piece of the entire dataset simultaneously, and the results will be consolidated on the outcome in a centralized grasp server. This will increase the calculated travel time. In this case, several research methodologies may be the same calculate. The total of the qualities of the instructional units that have been segmented Another option, which is solely based on map-lessen, is possible in a laptop dock with several centers or CPUs. Each CPU or center will analyze a section of the enormous information to deliver the best answer and accelerate the approach of AI researching procedures, suggesting a task inside the rapid computation. Since they are consecutively and use the strategies to a pattern parallel computing to enhance their effectiveness. Conversely, the authors provide a map-lessen technique for addressing and regulating the complexity of large amounts based on the respondent data. The authors confirm that it is an effective way of complying while spotting anomalies in website navigation occurrences for a big dataset. This works by dividing the data into subsets and reviewing each set for a simultaneous computation. Even when using a pattern to anticipate the bus's arrival rate. They make use of Grouping. K-Mean standards for allocating wandering time on a bus. The K-Mean set of factors can sort data into appropriate groups. It runs several times to identify several correct settings. The authors then use Back-propagated learning NN to predict the bus's arrival rate on a navigational network. Compared to a typical Backpropagation learning approach, the findings reveal that the predictions seem to be more correct and quicker. Similarly, utilize map-lessen to foresee visits to a Russian maritime institution's website gliding. On the other hand, the authors use the K-Nearest Procedure to examine glide's geographical function. Also, according to the rules, admission statistics are categorized mostly based on a democratic majority of its neighbors. The outputs are assigned to the least common sophistication among its Nearest neighbors. The greatest okay institutional selection is based on the witnessed, and the statistic's uncertainties may be lowered by charging a greater price for the type of difficulty.

9.4 FUTURE PREDICTED ARTIFICIAL INTELLIGENCE (AI)

Innovations with pattern recognition nonstop to disseminate the mystery at the heart of the vast amount of data collected in many sectors in 2016, in keeping with the scale of the marketplace for this method, become worth 272 million USD. It's far predicted to develop because of its excessive information garage capability, correct computing electricity, and cap potential to manipulate the complexity inside excessive extent information. This fee becomes primarily based on the utility of deep studying in photograph popularity duties for the medical industry and Instagram's face attractiveness traits. In addition, the manufacturing, financial, and knowledge discovery industries continue to improve their operations by implementing DL AI technologies. Furthermore, detecting patterns in data to make accurate predictions for the future would result in a rapid increase within the cost of deep researching by using 2025 to reach 10.2 USD billion. According to the author, the price of AI use may increase business, economics, and community for the transportation sector. They claimed that AI's current DL NB could provide a greater fee than traditional tactics. As shown in Table 9.1, the main areas where significant benefits may be obtained are travel and shipping-related industries. For example, AI tactics might be utilized to choose the best and shortest route for the convenience of highway consumers and cargo providers. One European company has managed to find real-time truck current effectiveness and operating force behavior by analyzing data ensuring that users. This contributed to a 15% reduction in gasoline cost and a reduction in shipment time. Furthermore, airlines can save cancelation costs by employing AI algorithms to forecast the weather and traffic bottlenecks. Other advantages include reduced visitor congestion, the most convenient public transportation schedule for customers, safeguarding passengers who use self-driving cars, and health and environment. On the whole, the charge for the trip region will rise from 62% to 128%, and the fee for transport and logistics will increase from 89% to 89%. In the future, the usefulness of autonomous automobiles will increase DL values.

9.5 FUTURE WORK

It has been cited that ANN is an extreme version because it may cowl a couple of AI duties and benefits also consist of inputs to outputs the usage of sample recognition. It can also manipulate many facts with proper adjustment and overall performance while surrounded with using noisy points. It saves time, considering it's for a quick computation device with correct overall performance. It is efficaciously carried out while examining distinct round systems and distinct duties. However, only a few studies have been carried out for a long-time period visitors country prediction the usage of deep studying structure. Also, preceding studies makes a specialty of simplest one or visitors' parameters to increase the version. As a result, future research may be devoted to strengthening predicting operations through larger than capacities and a couple of hidden nodes for the

TABLE 9.1 The Performance Improvement from AI [24]

Industries Using AI Techniques	Economic, Social, and Business Value (%)
Aerospace and defense	30
Advanced electronics/semiconductors	36
Insurance	38
Medical products	39
Telecommunications	44
Public and social sector	44
Healthcare systems and services	44
Banking	50
Agriculture	55
Consumer packed goods	55
Basic materials	56
Media and entertainment	57
Chemicals	67
Oil and gas	79
High tech	85
Automotive and assembly	85
Retail	87
Transport and logistics	89
Travel	128

model's structure. The objective is to forecast future traffic based entirely on historical and real-time data collected from detection stations located flow rate of the water and cameras attached to the highway. A Metropolitan throughway may be designated as a Proving Ground to illustrate the project's feasibility. Early in the studies, the particular site and knowledge gained could be discussed with throughway officials. As demonstrated in previous studies, the key to a successful installation of ANN is to choose the proper architecture for the community, i.e., the number of connections inside the number of hidden layers of convolutional networks. A fully convolutional device is used to identify the input information's more extensive and more dispersion capabilities (e.g., permutations of on-ramp and midline traffic flows, geometrical bottlenecks, and meteorological factors) when merged may produce specific traffic conditions on the freeway.

9.6 CONCLUSION

"This chapter provides an overview of AI systems for a variety of delivery-related difficulties. The number of functions is expected to increase as our cities and towns grow." Delivery structures end up extra instrumented, presenting much-wanted information for AI utility development. The evaluation centered on some of the utility regions predicted to have different effects on destiny towns, autonomous cars, public delivery, disruptive city mobility, computerized incident detection, destiny site visitors repute prediction, and site visitors control to manage. It indicates that AI may be used to resolve the growing journey call for Fuel combustion, environmental issues, and squandered fuels. The textbook is chock-full of case examples that show how AI may help with development. And growing a general populace for the neighborhood, locating world-class agenda plans for efficient public transport administration, improving review and input for site visitors signaling, and customizing itineraries for particular drivers. It is also employed for computerized reporting of incidents, identifying abnormalities during flights, and image multiprocessor sequencing for highway knowledge. Furthermore, in recent years, AI has been developed to use in site visitors demand prediction, the climatic change gave case forecasting, and future site visitors great nation for management and monitoring functions, as well as to relieve congested roads and quick decision making in the curriculum of risky scenarios, such as road accidents. It can also allow the government to determine either to develop new constructions or widen lanes; a choice on what way to proceed to take in the following of an event or extreme weather events; and a selection on the sum of funds associated with the preservation and reconstruction. Furthermore, personality vehicles and computerized public transportation systems rely on Advanced technologies to avoid inconvenience, mishaps, and overpopulation. Critical AI difficulties have also been resolved, most notably the idea that DL methods are "computational models" and the problem of bias in knowledge discovery introduced due to allowing individuals to recognize educational instances.

KEYWORDS

- **AI algorithms**
- **artificial intelligence**
- **cellular computation**
- **deep neural network**
- **knowledge base system**
- **nearest neighbor**

REFERENCES

1. Sadek, A., (2007). Artificial intelligence in transportation. *Transp. Res. Circ., E-C113*, 72–79.
2. Yegnanarayana, B., (1999). *Artificial Neural Networks* (p. 476). PHI Learning Pvt. Ltd.: New Delhi, India.
3. Abraham, A., (2005). Artificial neural networks. In: Sydenham, P. H., & Thorn, R., (eds.), *Handbook of Measuring System Design.* John Wiley & Sons, Ltd.: Hoboken, NJ, USA.
4. Minsky, M., & Papert, S., (1969). *Perceptron Expanded Edition.* MIT Press: Cambridge, MA, USA.
5. Kuldeep, S. K., Jagjit, S. D., & Krishan, K., (2021). Blockchain of IoT based earthquake alarming system in smart cities. *Book Entitled: Integration and Implementation of the Internet of Things Through Cloud Computing.* Published in IGI Global, ISBN 13: 9781799869818, ISBN10: 1799869814, EISBN14: 9781799869832.
6. Parker, D., (1985). *Learning Logic: Technical Report TR-87, Center for Computational Research in Economics and Management Science.* The MIT Press: Cambridge, MA, USA.
7. Getoor, B., & Taskar, L., (2007). *Introduction to Statistical Relational Learning; Volume L of Adaptive Computation and Machine Learning.* MIT Press: Cambridge, MA, USA.
8. Bacciu, A., Micheli, D., & Sperduti, A., (2012). Compositional generative mapping for tree-structured data—Part I: Bottom-up probabilistic modeling of trees. *IEEE Trans. Neural Netw. Learn. Syst., 23*, 1987–2002.
9. Krizhevsky, A., Sutskever, I., & Hinton, G. E., (2012). ImageNet classification with deep convolutional neural networks. *Adv. Neural Inf. Process. Syst., 60*, 1–9.
10. Kuldeep, S. K., Jagjit, S. D., & Santar, P. S., (2021). Blockchain technology for health care. *Book Entitled: Healthcare and Knowledge Management for Society 5.0: Trends, Issues, and Innovations.* Published in CRC Press, ISBN 9781003168638.

11. McCann, M. T., Jin, K. H., & Unser, M., (2017). Deep convolutional neural network for inverse problems in imaging. *IEEE Signal Process. Mag.*, *34*, 85–95.
12. Zhang, X. Y., Yin, F., Zhang, Y. M., Liu, C. L., & Bengio, Y., (2018). Drawing and recognizing Chinese characters with recurrent neural network. *IEEE Trans. Pattern Anal. Mach. Intell.*, *40*, 849–862.
13. Du, M. N. S., & Swamy, K. L., (2014). Recurrent neural networks. In: *Neural Networks and Statistical Learning*. Springer: London, UK.
14. Caterini, D. E., & Chang, A. L., (2018). Recurrent neural networks. In: *Deep Neural Networks in a Mathematical Framework*. Springer: Cham, Switzerland.
15. Jagjit, S. D., Kuldeep, S. K., & Amit, P., (2022). Implementation and deployment of 5G- drone setups. *Book Entitled: Internet of Drones: Opportunities and Challenges*. In Apple Academic Press (AAP), Canada, Hard ISBN: 9781774639856.
16. Australian Infrastructure Audit, (2015). *Infrastructure Australia*. Our infrastructure challenge. Available online: https://www.infrastructureaustralia.gov.au/listing/speech/speech-parliamentary-friends-better-cities-infrastructure-our-cities-challenges-ahead-philip-davies-ceo-infrastructure-australia (accessed on 20 January 2024).
17. *Linking Melbourne Authority and linking Melbourne*. Annual Report. Available online: https://www.parliament.vic.gov.au/file_uploads/Linking_Melbourne_Authority_Annual_Report_2014-2015_CwBGv8WN.pdf (accessed on 2 November 2023).
18. Klügl, F., Bazzan, A. L. C., & Ossowski, S., (2010). Agents in traffic and transportation. *Transp. Res. Part C Emerg. Technol.*, *18*, 69, 70.
19. Dogan, E., & Akgüngör, A. P., (2013). Forecasting highway casualties under the effect of railway development policy inˇ Turkey using artificial neural networks. *Neural Comput. Appl.*, *22*, 869–877.
20. Budalakoti, S., Srivastava, A. N., & Otey, M. E., (2009). Anomaly detection and diagnosis algorithms for discrete symbol sequences with applications to airline safety. *IEEE Trans. Syst. Man Cybern. Part C Appl. Rev.*, *39*, 101–113.
21. Akgüngör, A. P., & Dogˇan, E., (2009). An artificial intelligent approach to traffic accident estimation: Model development and application. *Transport*, *24*, 135–142.
22. Dia, H., & Rose, G., (1997). Development and evaluation of neural network freeway incident detection models using field data. *Transp. Res. Part C Emerg. Technol.*, *5*, 313–331.
23. Wang, R., Fan, S., & Work, D. B., (2016). Efficient multiple model particle filtering for joint traffic state estimation and incident detection. *Transp. Res. Part C Emerg. Technol.*, *71*, 521–537.
24. Wang, R., & Work, D. B., (2014). Interactive multiple model ensemble Kalman filter for traffic estimation and incident detection. In: *Proceedings of the 17th International IEEE Conference on Intelligent Transportation Systems (ITSC), Qingdao, China* (pp. 804–809).
25. Dia, H., (2001). An object-oriented neural network approach to short-term traffic forecasting. *Eur. J. Oper. Res.*, *131*, 253–261.
26. Huang, W., Song, G., Hong, H., & Xie, K., (2014). Deep architecture for traffic flow prediction: Deep belief networks with multitask learning. *IEEE Trans. Intell. Transp. Syst.*, *15*, 2191–2201.

27. Jiang, H., Zou, Y., Zhang, S., Tang, J., & Wang, Y., (2016). Short-term speed prediction using remote microwave sensor data: Machine learning versus statistical model. *Math. Probl. Eng.*, 9236156.
28. Ledoux, C., (1997). An urban traffic flow model integrating neural networks. *Transp. Res. Part C Emerg. Technol.*, *5*, 287–300.
29. Lv, Y., Duan, Y., Kang, W., Li, Z., & Wang, F. Y., (2014). Traffic flow prediction with big data: A deep learning approach. *IEEE Trans. Intell. Transp. Syst.*, *16*, 865–873.
30. More, R., Mugal, A., Rajgure, S., Adhao, R. B., & Pachghare, V. K., (2016). Road traffic prediction and congestion control using artificial neural networks. In: *Proceedings of the 2016 International Conference on Computing, Analytics and Security Trends (CAST)* (pp. 52–57). Pune, India.
31. Wu, Y., Tan, H., Qin, L., Ran, B., & Jiang, Z., (2018). A hybrid deep learning-based traffic flow prediction method and its understanding. *Transp. Res. Part C Emerg. Technol.*, *90*, 166–180.
32. Theofilatos, A., Yannis, G., Kopelias, P., & Papadimitriou, F., (2016). Predicting road accidents: A rare-events modeling approach. *Transp. Res. Procedia*, *14*, 3399–3405.
33. Król, A., (2016). The application of the artificial intelligence methods for planning of the development of the transportation network. *Transp. Res. Procedia*, *14*, 4532–4541.
34. Xu, T., Wei, H., & Wang, Z. D., (2009). Study on continuous network design problem using simulated annealing and genetic algorithm. *Expert Syst. Appl.*, *36 Pt 2*, 2735–2741.
35. Ceylan, H., & Bell, M. G. H., (2004). Traffic signal timing optimization based on genetic algorithm approach, including drivers' routing. *Transp. Res. Part B Methodol.*, *38*, 329–342.
36. Ulusoy, Ü., Sivrikaya, S., erifoglu, G., & Bilge, F., (1997). A genetic algorithm approach to the simultaneous scheduling of˜ machines and automated guided vehicles. *Comput. Oper. Res.*, *24*, 335–351.
37. Karoonsoontawong, A., & Waller, S., (2006). Dynamic continuous network design problem: Linear bilevel programming and metaheuristic approaches. *Transp. Res. Rec. J. Transp. Res. Board*, *1964*, 104–117.
38. Xu, T., Wei, H., & Hu, G., (2009). Study on continuous network design problem using simulated annealing and genetic algorithm. *Expert Syst. Appl.*, *36 Pt 1*, 1322–1328.
39. Kirkpatrick, S., Gelatt, C. D., & Vecchi, M. P., (1983). Optimization by simulated annealing. *Science*, *220*, 671–680.
40. Yaseen, S. G., & L-Slamy, N. A., (2008). Ant colony optimization. *Int. J. Comput. Sci. Netw. Secur.*, *8*, 351–357.

CHAPTER 10

Machine Learning and IoT for Urban Intelligence

PRIYA PROWAL,[1] NILANJANA PRADHAN,[2] SHRDDHA SAGAR,[3] and AJAY SHANKAR SINGH[3]

[1]PhD Scholar, School of Computer Science and Engineering, Galgotias University, Greater Noida, Uttar Pradesh, India

[2]Research Scholar, School of Computing Science and Engineering, Galgotias University, Delhi-NCR, India

[3]Professor, School of Computer Science and Engineering, Galgotias University, Greater Noida, Uttar Pradesh, India

ABSTRACT

Internet of things (IoT) is still at its nascent stage, but a portion of it emerges every day in our biosphere, which in turn influences different facets of our lives, even though we may not be aware of it. IoT is universal, whether it's manufacturing equipment or farm or grocery delivery in urban communities. Practical IoT implementations can be seen from the case studies already in operation. The agricultural sector has emerged as the major industrial-scale implementer of IoT. The case study of Tom Farms, which is a multigenerational family-owned farm in northern Indiana, was covered in an article in the New York Times. So far this is the most detailed case study to date. Smart thermostat, ML algorithms based open stack cluster control framework; smart lifting demo, optimization of air quality travel; smart lift demo and air quality optimization model using ML IoT; IoT based, smart hotel, IoT based Smart Home; cloud analytics;

Smart Mobility and Intelligent Transportation Systems for Commercial and Hazardous Vehicles.
T. Poongodi, Naga Swetha Pasupuleti, and Balamurugan Balusamy (Eds.)

cloud sleep tracking platform. However, over time, they all evolve. Several hospitals have been using IoT to incorporate a method of heart rate control. Another breakthrough in this area is fatty liver disease prediction using machine learning (ML) algorithms. To predict liver disease, a ML algorithm is used that could assist doctors in categorizing vital patients and developing an accurate diagnosis. Robotic Surgery is one of the most exciting technologies emerging in this area. In the area of medical surgery, robotic surgery is currently regarded as a groundbreaking discovery. The potential of IoT to optimize the manufacturing process from drug development to production has been realized by the medical system. Similarly, the Smart electronic health record program, which is a clinical management platform based on cloud technology operated by Amazon Web Service, is another innovative IoT application-based service (AWS). We will address the robotic surgery case study in this chapter that uses the combination of IoT and ML.

10.1 INTRODUCTION

The internet of things (IoT) is a phenomenon that represents an associated arrangement of everyone, everything, at any stage, any location, any assistance, and any system. The IoT is an emerging movement in front of the company and can have an impact on the company as a whole and can be seen as a link between strikingly unmistakable splendid articles and gadgets within the current network structure with extended extensions. Robotization has become feasible in each region, and IoT provides the answer for a wide range of uses. Traffic blocking, basic well-being, protection, crisis administration, coordination, retail, mechanical control, and robotization of human services are introduced in smart urban communities. The interested reader wants a better understanding of the IoT. In most desirable IoT application regions, a domain is occupied by clinical consideration and medicinal services [1]. The IoT can give rise to various clinical applications, such as remote health monitoring, fitness equipment, constant illnesses, and old consideration, for example. Another important possible application is the continuity of care and medication at home and through social insurance agencies. Therefore, different clinical gadgets, sensors, and symptomatic and imaging gadgets may be regarded as smart gadgets or objects that make up the IoT centerpiece. Administrations of

IoT-based human services depend on decreasing prices, rising personal satisfaction, and advancing the understanding of the client. The IoT will theoretically minimize gadget personal time by remote provision from the point of view of human services suppliers [1]. In addition, the IoT can effectively discern ideal occasions for their smooth and ceaseless operation for recharging supplies for various gadgets. In addition, by ensuring the best use and administration of more patients, the IoT accommodates the efficient booking of restricted properties. It has been shown that the use of artificial intelligence (AI) in medicinal services has been beneficial for as long as scarcely any year. A shortage of well-trained medical personnel is also present. This is where its significance can be found in machine learning (ML). It is possible to build a ML framework for health care services. This will prove to be a significant advancement because the diagnosis and treatment process can be re-engineered, and patients can benefit from reliable laboratory sample test results within a much shorter time period as well.

The IoT plays an important role in the healthcare sector. Many patients and individuals are unable to go through therapeutic therapy, and if there are innovative IoT technologies that can help to resolve this impairment and make their lives simpler, it will be important. The most recent models used for the healthcare system are IoT. IoT devices allow consumers to minimize the well-being that contributes to human services hazards and costs by gathering information and Investigation data from patients for the purpose of exchanging information and using distributed computing. A cloud-based platform was proposed in which different sensors were used to measure the different characteristics of the patient's circulatory system, heart rate, temperature, etc. In order to anticipate the data of the disease sensor and to transfer the information to the specialist via the cloud.

We interconnect a series of smart devices with each other through the internet on the IoT technology. The implementation of this framework uses approaches used to classify different health conditions and their predictions. The aim of such a structure is to rapidly estimate the disease by variables such as blood pressure, body temperature, and pulse rate of the patient. Patient data is accumulated with the aid of wireless communication sensors. Record patient data and provide emergency alerts, if necessary, using the IoT. It enables cloud storage of patient information. The historical record of the patient would therefore be beneficial to the doctors [3]. Timely action, therefore, guarantees favorable patient outcomes. In healthcare, data mining plays a significant role in disease prediction. To diagnose a disease, a patient's health test needs an N number. A range of assessments can be minimized by using data mining techniques. "The process of selecting, realizing, and displaying vast amounts of data to be tracked in encounters with useful hidden patterns is data mining technology" [4]. Using data mining problems using methods such as a spectrum of soft computing, ML, data visualization, and multidimensional databases, prediction of different types of diseases is systematically solved [4]. For the prediction of various diseases, different instruments and methods are used earlier in order to predict the disease. In hospitals, where patient expansion is regularly tracked by monitoring certain critical sensors that can be carried out by physicians, such as body temperature, pulse, and blood pressure. For just a few device prediction purposes, diagnostic data is usable. They are limited by a wide range of rules of association that implement the condition's diagnosis, depending on the intuition of the doctor and the record of the patient. We can track patients in real time by implementing this project. ARDUINO acts within the patient's body like a wireless sensor [5]. The health monitoring device for disease prediction at home is intended to support patients who live at home alone.

10.2 IOT AND HEALTHCARE

ML models can help simplify or accelerate tasks to make better health care decisions for patients and physicians. By helping to explain major data anomalies that are otherwise undetected, these approaches may produce better analysis for healthcare specialists. A few apps are as follows—using ML technology, and enhanced surgical robotics-surgery, like cardiac surgery, angioplasty requires a great deal of accuracy by surgeons and hospital personnel, because a lack of that contributes to life-dynamic effects for patients and is sometimes fatal. Complication risk prospects related to surgery are around 3–17%. In this way, if the prepared robots perform essential medical procedures, for example, the ones referenced above, it will gain more precision and movement right now. In the medicinal area, AI may end up being a desirable role included in the planning of errands and clinical data. The advantages of IoT devices will ultimately transcend surgical failure and decrease hospital costs [6].

Innovative techniques such as robotic suturing require rigorous testing and preparation systems to create a degree of precession [6] for the machines. The cost of educating specialists must also be taken into account in order to make them knowledgeable in the use of robots as a part of their surgical training [6]. The absence of data and knowledge about this makes it difficult to determine the appropriate time range. ML flourishes as large amounts are used for pattern recognition. The aim should therefore be to use AI to make robots that are safe for individuals, but stable and versatile to wander in an unpredictable domain, on the other hand.

10.3 HEALTHCARE IOT APPLICATIONS

IoT is useful in the area of healthcare. Different applications have been developed that are useful for patients and physicians. New applications currently available on the market are being built day by day. The numerous improved healthcare solutions include IoT advances in healthcare. With the IoT system, emerging developments in healthcare systems and general pharmacy practice are expanded. The reach of experts inside a facility is rising. The various data obtained from enormous real case arrangements improve both the precision and scale of medicinal data. Additionally, the accuracy of healthcare delivery is strengthened by integrating increasingly dynamic developments in the medical sector [7].

Various applications based on IoT are:

- Kit for glucose level sensing;
- Monitoring blood pressure;
- Monitoring electro-cardiogram;
- Charting for health care;
- Remote tracking.

10.4 MACHINE LEARNING (ML) AND HEALTHCARE

Elements of a person's well-being should be observed and treated with the most serious treatment. With proactive verification of human health, numerous diseases can be minimized. As many wearable gadgets and health tracking systems are available on the market, creative progress is at its height. Also, experts are of the view that the medical conditions from the signs observed are very difficult to foresee. Two advanced scientific methods, such as IoT machines, have made it much easier for physicians to identify the root cause of diseases and predict their severity using modern algorithms along with artificial intelligence (AI) and **ML**. **ML** algorithms are used in this study to observe human health problems. The training and testing using various **ML** algorithms [8] were initially applied to the UCI dataset. Using IoT devices, the patient's blood pressure and temperature are collected using IoT devices in the test process. The person's health care symptoms are calculated from the sensor data obtained in the test process from the IoT devices. In order to gauge the prediction percentage accuracy, statistical analysis is performed on the data obtained from the cloud via IoT devices. The findings obtained from the K-nearest Neighbor algorithm and other techniques of classification are also used here.

The central aspect of human life is health care. Continuous development and maintenance in the healthcare sector are increased by the creative invention of IoT devices.

In order to have the help of demonstrative devices such as CT, MRI, PET, SPECT, and so on, anomalies available well below the skin can be determined. In addition, it is possible to detect such anomalous conditions before their occurrences, such as heart attack, and epilepsy [9]. The steady rise in the population and also a flighty spread of endless disease among the masses have put a strain on the modern medicinal system, and demand for services from hospitals to doctors and nurses is extremely high. For patients with diabetes, or for patients with another particular disease such as Perkinson's Disease, previous research has already dealt with the application of IoT. There is a significant opportunity for research in the field of emergency health care services to be promoted. Several other related works were thoroughly analyzed and surveyed.

ML is an area of AI that emerges from. We can assemble stronger, more informative machines by implementing ML. ML is a technique to be observed from experience, without being directly altered. The data is placed directly to the algorithm instead of writing code, and the logic is constructed based on the data given. AI is used in web searches, filtering of spam, promotion agreements, exchange of stocks, etc. ML increases the relevance and understanding of massive information and distributed computing by analyzing these enormous parts and disentangling the assignment of data scientists in a robotic procedure. In various regions, from the design of sciences to relational partnerships, industry, biomedical research, and protection, far-reaching educational assortments are assembled and analyzed.

The IoT architecture offers a forum for disease detection by using a sensor network to observe and guide patients. Both hardware and software are used in IoT appliances. As hardware, the heart rate monitor, the temperature sensor, the blood pressure sensor and the Raspberry Pi board [10] are included. The method steps are as follows: collecting data stored in the cloud using the sensor and analyzing the data to verify the person's health status (Figure 10.1).

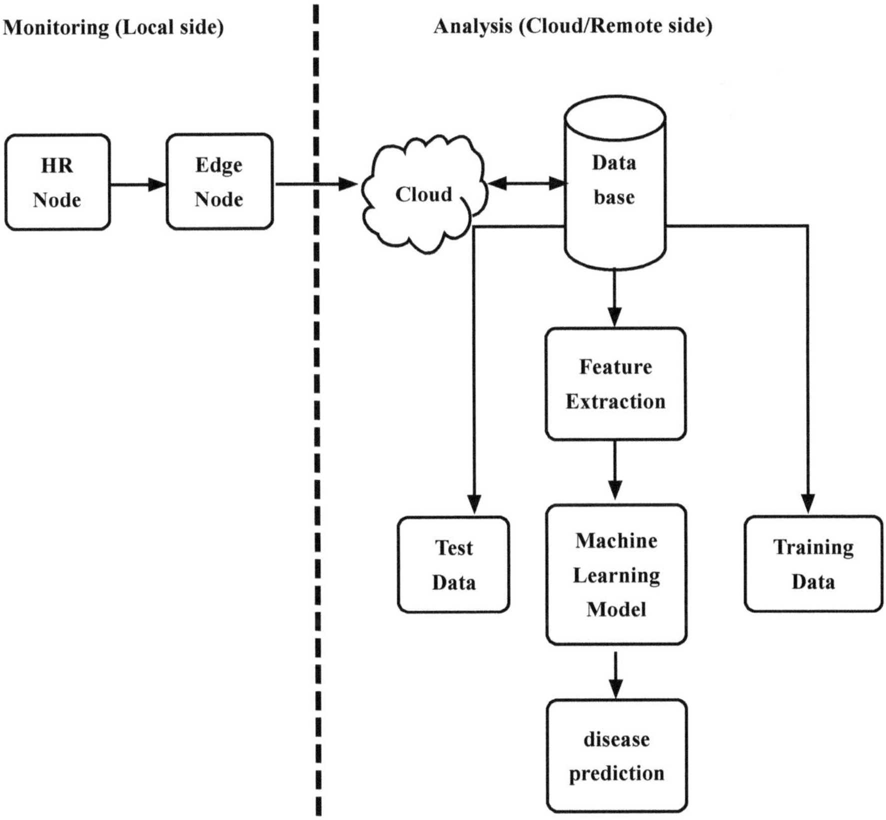

FIGURE 10.1 Disease prediction model based on machine learning.

Differences arise when a surgical procedure is not performed on a body part. A seizure in the brain will cause the heart rate to increase. To determine the heart rate at any moment, the sensors monitor the heartbeat. In order to accurately record the sensor's characteristics, the sensor is connected to the Raspberry Pi board. Using the serial monitor or the LCD display interface to demonstrate the values.

10.5 HEALTH STATE PREDICTION BY MACHINE LEARNING (ML) ALGORITHMS

For analysis of the dataset, a **ML** approach can be used. The classification methods were used to construct the classification model from the data set given. Classifications such as the decision tree and neural network, KNN (Naive Bayes), the supporting vector machine, etc., are used in various typing techniques.

Any method for detecting a model that best suits the relation between the class label and the data set attribute is used with the **ML** algorithm. The model constructed using an educational algorithm could adjust input data to the class labels of documents it had not properly observed. The objective is to construct specific models for each learning algorithm. Using the **ML** algorithm provides training on the basis of data for medical system analysis and training. Classification methods for K-Nearest neighbor are used [11].

10.6 MACHINE LEARNING (ML) APPLICATIONS IN PHARMA AND MEDICINE

The fluctuation in the growth of AI materiality in medical care enables us to predict a future in which hip-related science and creativity benefits countless patients, even without awareness. ML-based applications will soon be found updated with constant patient data, made available in various countries from different healthcare systems, improving the productivity of previously inaccessible new therapy decisions [12].

These are the top 10 wellness ML applications:

1. **Diseases and Diagnosis Recognition:** The determination of diseases and illnesses, which are in any case considered difficult to examine, is one of the key ML applications of medicinal services. This can include everything from malignant growths that are difficult to discover, to other genetic disorders during the underlying stages.

In medical fields, for example, oncology, the biopharma giant is using AI to invent therapeutic therapies. PReDicT (predicting response to depression treatment) from P1 vital means developing a monetarily realistic approach to evaluating and delivering treatment under normal clinical conditions [13].

2. **Drug Discovery and Creation:** One of the basic clinical applications of AI is the sedate exposure measure in the early phase. This also involves developments in R&D, such as cutting-edge sequencing and precision solutions that can assist in discovering elective ways to treat multi-factorial diseases. Currently, ML procedures involve unsupervised learning that, without giving any predictions, can recognize patterns in data. Microsoft's Hanover business uses ML-based technologies for various activities, including the development of AI-based disease treatment innovations and the customization of the AML (acute myeloid leukemia) drug mix [14].
3. **Diagnosis with Diagnostic Imaging:** Both deep learning (DL) and ML are responsible for the breakthrough called Computer Vision for achievement. This has been recognized in the Microsoft-developed InnerEye operation that operates using image diagnostic tools for image processing. As ML becomes more and more open and grows in its illustrative limit, expect to see more sources of knowledge from varied medicinal symbolism becoming a part of this demonstrative AI-driven procedure [15].
4. **Customized Medicine:** In addition to blending singular well-being with the prescient test is also ready for more exploration and better infection assessment, personalized medicines may be more persuasive. Specialists are currently restricted to looking at a specific set of analyzes or estimating the patient's risk based on his symptomatic background and available genetic data. ML in drugs is taking enormous strides in any case, and IBM Watson Oncology is a pioneer in this growth by leveraging ongoing restorative history to create new treatment options. In future we will see more devices and biosensors with state-of-the-art well-being estimation capabilities, which allow more data for these ML-based technologies to be readily available from the frontline.

5. **Behavioral Change based on Machine Learning:** Behavioral modification is an integral aspect of preventive medicine, and numerous new companies are developing in the fields of cancer detection in the early stage of the multiplication of ML in the medical sector.

And the identification of other diseases. Somatix is a data analytics company focused on B2B2C that has developed an ML-based application to interpret signals that we make in our everyday lives, enabling us to understand our negligent actions and introduce fundamental improvements.

10.7 PREDICTION FOR OUTBREAK

IBM Watson Oncology pioneered this development with the use of a continuous history of restoration to build new treatment options. Further devices and bio-sensors can be seen in the future with state-of-the-art well-being estimates that allow additional data to be available from the front line for certain ML-based technologies. In underdeveloped countries, predicting these outbreaks is especially helpful because they need essential medical foundations and instructive structures. The ProMED-mail, an internet-based revealing stage that screens emerging diseases and rising ones and constantly provides outbreak updates, is an important case of this.

The framework suggested: collecting and sharing information with each other on IoT devices makes gathering information feasible and producing data more specifically in order to build even another data stream quickly. It foresees a disease dependent on IoT, and, at an early stage, the analytical system predicts the disease. Most people affected by illness suffer because, at the right time, they don't get proper care. Patients' existing health records are obtained using sensors. A sensor is a device of physical data being tested and interpreted and then a signal is sent to the controller. Sensors receive knowledge and resolve human error from the human body. The ARDUINO microcontroller acts as a bridge to the wireless sensor which, for analysis and storage, is sent to the cloud.

The IoT Healthcare System screens the patient's well-being status, sending the cloud and cloud that holds patient data on its servers to the ongoing patient health status. The aim of the prediction system for IoT diseases is to give the individuals concerned the most advantageous data. As suggested by the well-being announcement, the professional will design the patient without much of a stretch. In this way, patients with illnesses will use the framework as soon as possible to reach a specialist.

A new area of study remains the IoT, and its dormant use of medical services resources is a region that is still in its infancy. In different kinds of fields, the web of things is used to improve the system of different things on a scale that used to be amazing. IoT is currently linking devices to each other for a day, through which living from human origins becomes simple. The heart is an integral part of the human body. If the heart does not function properly, it can stimulate different coronary disorders and affect different body parts in addition. Coronary disease is one of the most lethal diseases, achieving 12 million annual expenditures. The effect of coronary disease is so unexpected that it causes us to pursue a scenario time and time again. In light of coronary disease in India, one person passes on at customary intervals. The IoT is investigated, and its social security awareness is included. A few leading developments are discussed in the construction of IoT systems for human administration. A non-exclusive and controlled model for the future all the way to IoT restorative administration structures is proposed, building on the regular themes of these works. Research in related fields has shown that remote control of prosperity is possible, but maybe the preferences it might offer in different settings are increasingly important.

Remote prosperity observation removes the need for a case wellspring of profit to check the pressure on numerous medical clinic resources right now for its own prosperity. Basically, it can increase access to social care assets without reducing the efforts of medical services frameworks and can continually give people greater authority over their own well-being. In today's world, heart diseases are the deadliest. In a split second, the heart disorders strike so that there is less time for a valid conclusion. The most troublesome errand for the therapeutic offices is the exact diagnosis of patients on a time-to-time premise. In general, the vast majority of crews can pass on information containing a lot of details, and after a specific measure of damage is done, various infections are distinguished in this way. If the various well-being criteria that are conceivable with the help of remote checking are observed, this can be significantly reduced. Remote observation involves different devices such as versatile software, wearable gadgets, etc. We use the IoT and knowledge mining in order to predict coronary disease by using sensors, such as ECG Sensors, i.e., AD8232 (Figure 10.2).

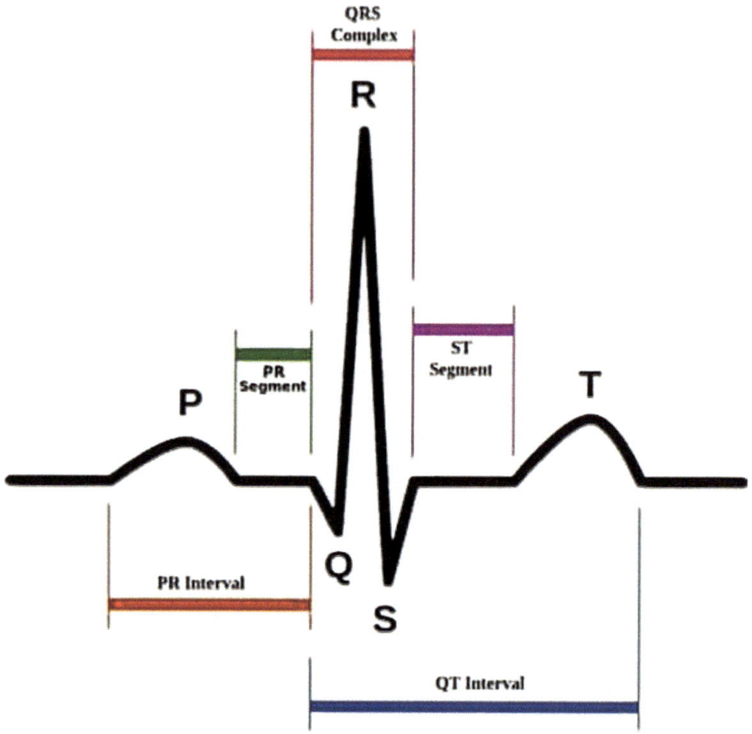

FIGURE 10.2 ECG wave.

Source: Public domain

The cardiac beat sensor comprises a diode and a light-emitting photodiode. In the front of the sensor, the LED is mounted. It estimates blood flow change in every organ in the body, which contributes to a change in light intensity in the organ.

So, the heart pulsates through the blood progression screen. After some time, the light will be analyzed to calculate heart thumps. At +5V or 3.3V, this heartbeat sensor can run. The ARDUINO AT mega328 microcontroller is open source. Without any issues, the AT mega328 microcontroller can be programmed, eradicated, and rebuilt. Since the software only needs a USB connection, ARDUINO is better than other controllers. Arduino's programming is regarded as open source. There is no provision for the consumption of code in IC by another gadget. The working voltage is 5V for the ARDUINO. It has 14 advanced information pins, 6 basic information pins, and yield pins. ARDUINO is responsible for monitoring all gadgets and gathering input data from sensors. In this sense, ARDUINO is an extension of remote sensors mounted in the body of the patient, such as a pulse sensor, a temperature sensor, a heartbeat rate sensor, and so on.

Use DL to boost mammography screening detection for breast cancer. DL is a part of AI techniques and is designed to tackle clinical imaging problems. Profound learning calculation promotes the use of a start-to-finish preparation method that can reliably classify bosom disease on screening mammograms and can skillfully use the preparation of datasets using either all clinical clarity or only the entire image's malignancy status. This approach needs wretched explanations only in the underlying preparation level, while the resulting stag. It was noticed that in order to classify screening mammograms, the coevolutionary network method performed in a way that is better than all distinct past techniques. A bunch of digitized film mammograms obtained from the Digital Screening Mammography Database (CBIS-DDSM) in order to carry out a free test, the best single model achieved an AUC of 0.88 for each image, and the usual four-model AUC improved to 0.91 (affectability: 86.1%, specificity: 80.1%), while in an autonomous full-field advanced mammography test collection (FFDM) images Such advances indicate that computerized DL techniques can be easily prepared to ensure high precision in heterogeneous mammography phases, and thus have a huge future for enhancing clinical administration by reducing bogus positive and bogus negative mammography screening performance.

In using these methods to enhance the accuracy of malignancy screening results, the advancement in AI procedures, particularly in DL, has fueled the revenue of the clinical imaging network. The bosom disease has emerged as the subsequent driving reason for malignancy transitions among U.S. ladies. This highlights the importance of mammography screening, which has been shown to minimize the death rate. Despite the expressed benefits of mammography screening, it is associated with a high probability of bogus positives as well as bogus negatives. In the U.S., computerized screening mammography has a typical affectability of 86.9%, while the normal specificity is 88.9%. PC aided recognition and analysis CAD programming have been in clinical use since the 1990s to assist radiologists in enhancing the prescient accuracy of screening mammography.

Nevertheless, the data shows that early market CAD systems have not triggered any vital implementation enhancement and development has declined since its presentation over 10 years. There has been a great deal of income in increasing DL devices to assist radiologists by enhancing the accuracy of screening mammography with the achievement of DL processes in visual object recognition and detection and in various different fields. Late studies have shown that a DL-based CAD system not only worked equally well as the independent mode of radiologists but also enhanced the presentation of the radiologists in support mode.

10.8 AUTOMATIC IOT SYSTEM USING DEEP LEARNING (DL) IN ROBOTIC SURGERY

Research in the application of autonomy vows to shift medical practice to intercession that is more stable, increasingly predictable, and insignificantly intrusive. When in transit to robot-helped systems, new developments continue and drive towards a future of entirely self-sufficient autonomous specialists. The da Vinci robot, whose friendliness has only recently been demonstrated in gynecology, urology, and overall medical care with remote-controlled laparoscopic treatment, is the most comprehensive operating system. DL is an advanced form of **ML** that improves the many challenges in the healthcare field. Clinical imaging, for instance in the area of breast cancer prediction, bone prediction, age analysis, and other problems has been successfully used as a neural network (CNN). DL apps for automated instrument segmentation have proven to be seriously executed in binary segmentation and promising results in the multi-class division. Deep neural network (DNN) changes, like clinical robots, are beginning to develop, and are suitable for use in portable and embedded gadgets.

The robotic surgical system has been expanding its role in medical surgery for the last two decades. Despite the Da Vinci Surgical System (Intuitive Surgical, Sunnyvale, CA) being the most recognized stage, several different devices are used for abdominal, urological, pelvic, cardiovascular, and neurological medical procedures. Robot-helped approaches deliver numerous favorable conditions over standard cautious procedures, such as improved accuracy, decreased tremor, and increased motion duration. It is not shocking that we are in the midst of a transition in perspective towards the coordination of the automated development of traditional surgical techniques.

i. **PISIoT:** A smart health platform for overweight and obesity management ML and IoT-based.

ii. **PISIoT:** Functionality and architecture.

PSIoT offers a consistent definition of the exercises and components of each module. This allows high adaptability to be taken into account. The PISIoT structure consists of five layers in Figure 10.3: 'the presentation layer, network control layer, integration layer, data management layer, and IoT-based services layer.' With specific usefulness and relationship, each layer is made of different pieces. Below is a generic representation of the layers:

FIGURE 10.3 Heartbeat sensor.

1. **Presentation Layer:** The patient and the stage impart through this layer, promising a web application and a convenient application in which patients can note and monitor their biomedical components, open IoT-based organizations, clinical history, and suggestions. The stage grants manual engagement of water and food during the day in expansion. Calories, carbohydrates, fat, and proteins are subsequently collected from the knowledge base of the system provider when patients access food or drinks used using a cell phone or PC with an Internet link to the wearable gadgets. However, in PISIoT, despite any real creation undertaken by the patient, it is possible to join food, or a reward avoided from the provider's database.

2. **Network Layer Monitoring:** This layer includes distinctive IoT-based products, such as wearable gadgets linked to media

transmission gear, which have correspondence interfaces that energize information exchange, for example. These gadgets allow information on the biomedical components of the patient (beat, calories consumed, rest, minutes of actual development, and weight) to be accumulated, taking into account the activities carried out during a day. For further preparation and review, all the information assembled is shipped from the data layer of the heads.

3. **Integration Layer:** This layer includes wearable and intelligent system manufacturers, and the patient provides the patient with information and advice to reply. Furthermore, this layer is responsible for listing IoT administrations together according to PISIoT recommendations.
4. **Data Analysis Layer:** This layer defines and develops therapeutic proposals for fundamental factors. This layer also provides valuable PISIoT and patient clinical history data.
5. **IoT-based Service Layer:** It is used to link, summon, pick, and confirm IoT-based service accessibility.
6. **Data Management Layer:** It is used to monitor and optimize the patient's past clinical history and the data gathered by the equipment used. Monitoring of the network layer consists of multiple IoT devices, such as portable and smart devices that have correspondence interfaces, such as data trading associated with media transmission hardware. These tools allow data on biomedical factors to be collected (pulse, calories consumed, rest, physical activity minutes and weight).
7. **Network Layer Monitoring:** It consists of various IoT devices, such as wearable and smart devices that have correspondence interfaces that, for example, enable data trading with media transmission hardware. These instruments allow the collection of data on the patient's biomedical factors (pulses, calories taken, rest, minutes of physical activity and weights) in view of daily activities. For further analysis and investigation, all data collected is sent to the data management layer (Figure 10.4).

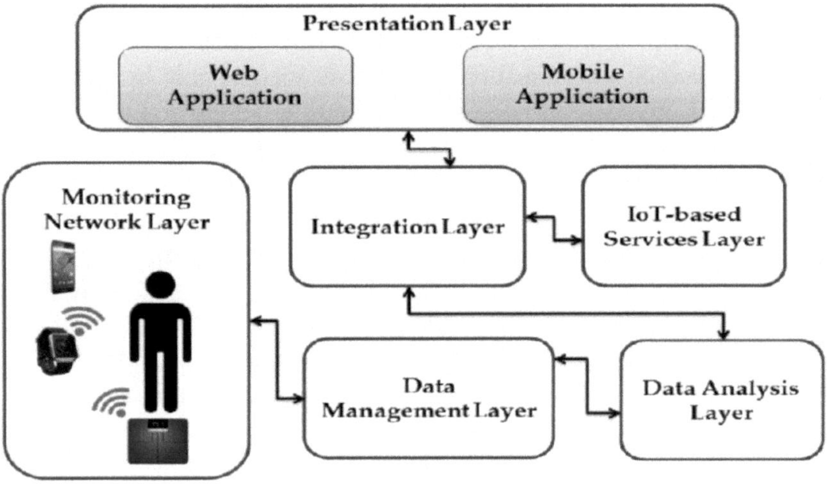

FIGURE 10.4 General architecture PSIoT.

10.9 CASE STUDIES

- **Case Study I:** This section is a case study that approves PISIoT's commitment to weight to control weight or misfortune in older patients who are increasingly inclined to localized myocardial necrosis with overwhelming weight problems. Biomedical factors in corpulent elderly patients were regulated in the sense of contextual research to restrict the amount of weight and thus reduce the risk of dead myocardial tissue. The situation is as follows: elders with heaviness have to know how many calories they eat, how many calories they drink, how to relax, what pulse, how much water they need to measure, how daily they take steps, and what they need to do to slowly lose weight without problems with their well-being. CVDs are dispersions of the heart and vein, including coronary disease, stroke, RCD, myocardial tissue, and other conditions as shown by WHO. Cardiovascular diseases (CVDs) are a symptom of heart and vein dispersion.
- **Case Study II:** CVD risk individuals have shown high pulse, glucose, and overweight lipids. CVDs' risk and proper care can prevent unintended losses, but CVDs continue to cause multiple

transfers worldwide. Figure 10.5 demonstrates the cases of elderly persons with heaviness who have side effects of myocardial dead tissue or who are supposed to lose weight only by getting localized myocardial necrosis. A wearable device and a skillful scale linked and synced to mobiles gather biomedical factors (pulses, rest, calories, weight, and physical activity minutes) and many factors (steps, floors, calories consumed, separation traveled, water expended, and worked out). Information can be gathered by mobile phone linked to any patient using wearable equipment and keen information from gadget suppliers. Different medical tests may also be undertaken in order to detect conceivable essential causes, establish recommendations and submit additional IoT (clinical studies) administration to achieve the patient's objective.

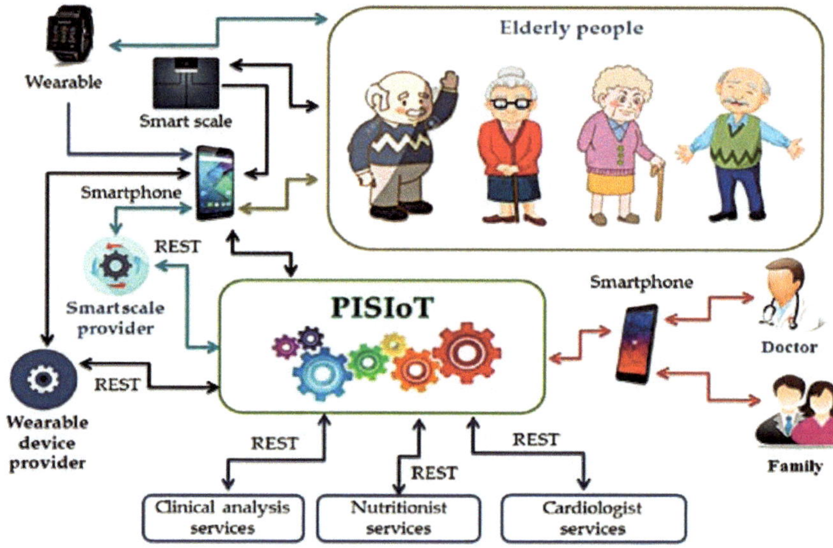

FIGURE 10.5 Scenario: Monitoring elderly people with obesity.

10.10 CONCLUSION

The human body constantly offers data on the state of well-being. This data is collected through structures or devices that calculate or classify the characteristics and factors of the body in an intrusive or non-intrusive manner for explicit purposes. Human services workers use biomedical factor estimates to decide on decisions on determinations and medications in order to enhance the well-being of patients. Personal satisfaction, particularly among the elderly, is significantly affected by overweight worldwide. In order to achieve a typical goal, the IoT makes it conceivable to interconnect, discern, acknowledge, and process information between items or administrations. Observing, researching, identifying, and regulating the diseases, such as overweight and heftiness, along with the age of recommendations to avoid them, are the main focus points of IoT in medicinal services. Be that as it may, there are minimal assets in the products used in the IoT, so it is necessary to think of various options for knowledge investigation, such as **ML**, for example. **ML** is one of the sub-parts of AI that, in the light of examples and derivations, involves considering the calculations and factual models used in PC systems to achieve explicit goals successfully.

KEYWORDS

- **cloud analytics**
- **deep learning**
- **internet of things**
- **machine learning**
- **open stack clustering**
- **surgical robotics**
- **urban intelligence**

REFERENCES

1. Islam, S. R., Kwak, D., Kabir, M. H., Hossain, M., & Kwak, K. S., (2018). The internet of things for healthcare: A comprehensive survey. *IEEE Access, 3*, 678–708.
2. Tuyishimire, E., & Bagula, B. A. (2020). Modelling and analysis of interference diffusion in the internet of things: An epidemic model. In *2020 Conference on Information Communications Technology and Society (ICTAS)* (pp. 1–6). IEEE.
3. Yang, Y., Wu, L., Yin, G., Li, L., & Zhao, H., (2017). A survey on security and privacy issues in internet-of-things. *IEEE Internet of Things Journal*, *4*(5), 1250–1258.

4. Hu, J., Perer, A., & Wang, F., (2016). Data driven analytics for personalized healthcare. In: *Healthcare Information Management Systems* (pp. 529–554). Springer, Cham.
5. Wai, K. T., Aung, N. P., & Htay, L. L. (2019). Internet of things (IoT) based healthcare monitoring system using NodeMCU and Arduino UNO. *Published in International Journal of Trend in Scientific Research and Development (IJTSRD)*, *3*(5), 755–759.
6. Azizian, M., McDowall, I., & Sorger, J., (2018). Visualization in robotic surgery. In: *The SAGES Atlas of Robotic Surgery* (pp. 53–65). Springer, Cham.
7. Ansari, S., Aslam, T., Poncela, J., Otero, P., & Ansari, A., (2020). Internet of things-based healthcare applications. In: *IoT Architectures, Models, and Platforms for Smart City Applications* (pp. 1–28). IGI Global.
8. Fatima, M., & Pasha, M., (2017). Survey of machine learning algorithms for disease diagnostic. *Journal of Intelligent Learning Systems and Applications*, *9*(1), 1.
9. Qadri, Y. A., Nauman, A., Zikria, Y. B., Vasilakos, A. V., & Kim, S. W. (2020). The future of healthcare internet of things: a survey of emerging technologies. *IEEE Communications Surveys & Tutorials*, *22*(2), 1121–1167.
10. Rohit, S. L., & Tank, B. V., (2018). IoT based health monitoring system using raspberry PI-review. In: *2018 Second International Conference on Inventive Communication and Computational Technologies (ICICCT)* (pp. 997–1002). IEEE.
11. Chen, M., Hao, Y., Hwang, K., Wang, L., & Wang, L., (2017). Disease prediction by machine learning over big data from healthcare communities. *IEEE Access*, *5*, 8869–8879.
12. Vamathevan, J., Clark, D., Czodrowski, P., Dunham, I., Ferran, E., Lee, G., & Zhao, S., (2019). Applications of machine learning in drug discovery and development. *Nature Reviews Drug Discovery*, *18*(6), 463–477.
13. Kingslake, J., Dias, R., Dawson, G. R., Simon, J., Goodwin, G. M., Harmer, C. J., & Ruhé, H. G., (2017). The effects of using the PReDicT Test to guide the antidepressant treatment of depressed patients: Study protocol for a randomized controlled trial. *Trials*, *18*(1), 558.
14. Kalaiselvi, K., & Karthika, D., (2020). Identifying diseases and diagnosis using machine learning. In: *Machine Learning with Health Care Perspective* (pp. 391–415). Springer, Cham.
15. Giger, M. L., (2018). Machine learning in medical imaging. *Journal of the American College of Radiology*, *15*(3), 512–520.
16. Machorro-Cano, I., Alor-Hernández, G., Paredes-Valverde, M. A., Ramos-Deonati, U., Sánchez-Cervantes, J. L., & Rodríguez-Mazahua, L., (2019). PISIoT: A machine learning and IoT-based smart health platform for overweight and obesity control. *Applied Sciences*, *9*(15), 3037.

Index

J

K

L

M

N

O

P

Q

R

S